RENEWALS: 691-4574
DATE DUE
MAY 12
NOV 21
APR 11
APR 21 2008
WITHDRAWN
UTSA LIBRARIES

Coal and Energy
The need to exploit the world's most
abundant fossil fuel

Derek Ezra

Coal and Energy

The need to exploit the world's most abundant fossil fuel

A HALSTED PRESS BOOK

JOHN WILEY & SONS
New York

First published 1978 by Ernest Benn Limited
25 New Street Square, Fleet Street, London EC4A 3JA
and Sovereign Way, Tonbridge, Kent TN9 1RW
Distributed in Canada by
The General Publishing Company Limited, Toronto

Published in the U.S.A.
by Halsted Press, a Division of
John Wiley & Sons, Inc., New York.

Printed in Great Britain
by Lewis Reprints Ltd., Tonbridge.

ISBN 0 470 26339-3
Library of Congress Catalogue Card No. 78-5785

Library of Congress Cataloging in Publication Data

Ezra, Sir Derek
Coal and Energy
1. Coal 2. Power Resources
I. Title
TN800. E97-333.8,2. 78-5785.

Contents

List of Illustrations

MAPS

PLATES (*between pages 94 and 95*)

Preface

This is a book about the energy problem, with particular reference to coal. It is about the world energy problem, with particular reference to Britain.

My reasons for writing it (with much help from others) are twofold.

First, the energy debate is in full spate, both here and elsewhere. If one has anything to contribute, now is the time to do it. The theme of this book is that we are fast getting to the stage when the great energy debate must be concluded, at least in its main lines. Action must be initiated now, world-wide, to avert the prospect of a real energy crisis within the next couple of decades.

My second reason is personal. I have been closely involved in the British coal industry for over thirty years, indeed for virtually the whole of the post-war period. I have had much to do with energy developments in Britain, in Europe and more widely. I am spending a large part of my time now in planning and implementing new strategies for coal within the overall energy context. Before I give up this role, I feel it may be of interest to record my views of the recent past and of future prospects.

What has happened in the British coal industry since the war mirrors the wider world energy situation. Under the first two chairmen of the National Coal Board, Lord Hyndley (1947–52) and Sir Hubert Houldsworth (1952–56), the problem was to produce as much coal as possible to meet the urgent needs of post-war reconstruction. This was succeeded by a totally different situation under the next two chairmen, Sir James Bowman (1956–61) and Lord Robens (1961–71).

It was then that oil in increasing quantity and at low prices made its appearance on the world's markets. Additionally, nuclear power was being pioneered in Britain and later on Dutch gas, then North Sea gas and oil were being discovered.

All this posed a severe threat for the relatively high cost coalfields of Western Europe. In the new circumstances of fuel abundance at low prices, governmental policy, both here and on the Continent, was to reduce the size of the coal industries. Indeed, in Holland, the objective was to close down the industry altogether, and this has since been done.

Many warning voices were raised about the risk of projecting too far ahead what might be (and indeed proved to be) a short term situation. But these voices – which included those of Lord Robens, his Board and the mining Trade Unions – were ignored. And so began a period of substantial and painful contraction for Britain's coal industry, skilfully and responsibly handled by close understanding between management and unions.

When I succeeded Lord Robens as Chairman of the Coal Board in 1971, two factors began to emerge which were due to have a profound effect on the next phase of the coal industry's long history. The first was a growing unsettlement within the National Union of Mineworkers, resulting partly from the sense of insecurity caused by the previous policy of contraction and partly from external factors then prevalent. This culminated in the two strikes of early 1972 and early 1974. The second factor was the new prospect for coal opened up by the massive increases in the price of oil in 1973 and 1974. Thus, for the third time since the last war a major twist was given to the affairs of the coal industry.

This book is concerned with this aspect, seen against the background of the world energy situation as it has emerged over the past quarter of a century and as it may develop over the next quarter of a century. It is an issue which goes considerably beyond the interests only of the British coal industry. Since 1973 there has been a rediscovery throughout the world of the importance of energy, and there is a realisation of its massive economic, social and political significance.

However uncertain the future may be, it is now of vital importance to make an assessment of forward energy trends and to take such action as may be necessary to avert within the next decade or so what could well be a far more serious world energy crisis than was experienced in 1973 and 1974. In this connection I believe that the role of coal, as the most abundant fossil fuel, will be crucial.

I have been helped by many friends and associates in the preparation of this book. Within the Coal Board I would like to express particular appreciation to Ian Forster, who has been our Energy Consultant for 5 years, and was previously Director of Economics and Statistics in what is now the Department of Energy. I would also like to thank Mike Parker, Director of Central Planning at the NCB and Colin Ambler who is in charge of my private office. Finally, my gratitude goes to Henry Bailey-King, Chairman of Charles Knight & Co. Ltd., a subsidiary of Ernest Benn Ltd. the publishers of this book. He is a friend of long standing, and without his encouragement and enthusiasm the book could not have been written.

Map 1

COALFIELDS OF GREAT BRITAIN

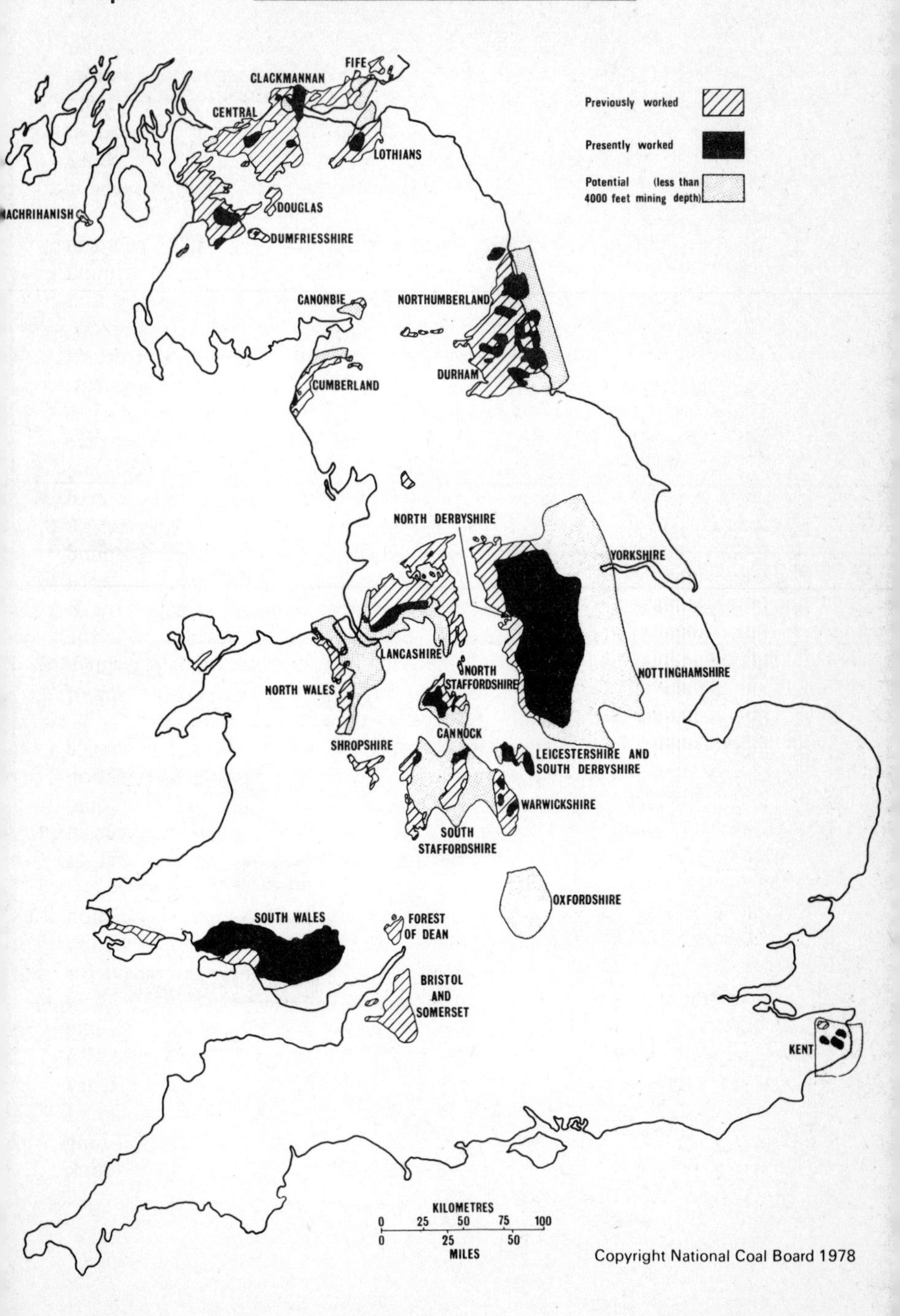

Map 2

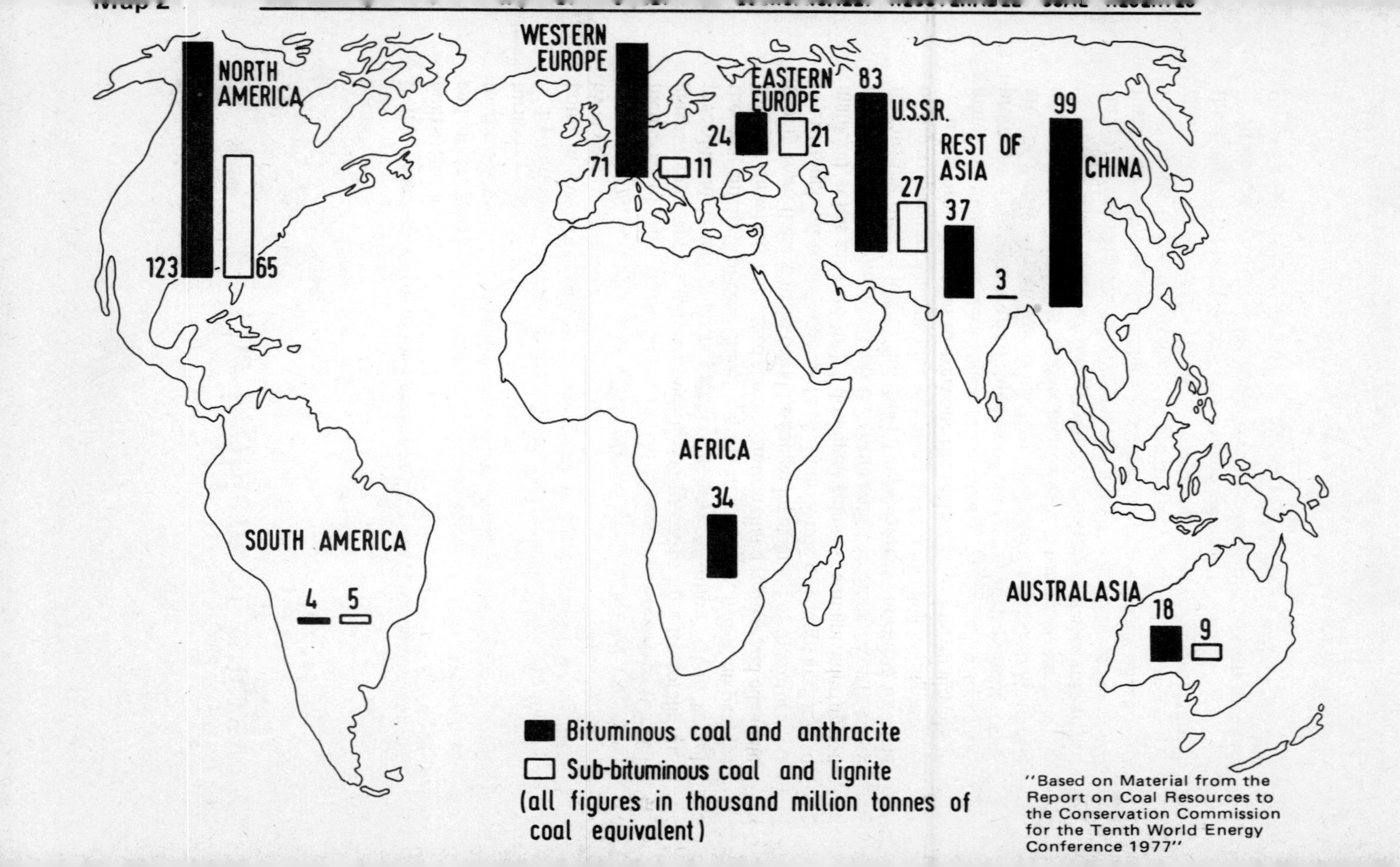
NORTH AMERICA
123
65
WESTERN EUROPE
71
11
EASTERN EUROPE
24
21
U.S.S.R.
83
27
REST OF ASIA
37
3
CHINA
99
SOUTH AMERICA
4
5
AFRICA
34
AUSTRALASIA
18
9
Bituminous coal and anthracite
Sub-bituminous coal and lignite
(all figures in thousand million tonnes of coal equivalent)
"Based on Material from the Report on Coal Resources to the Conservation Commission for the Tenth World Energy Conference 1977"

1 The Problem of Energy

It has been remarked more than once that the pronouncements of politicians, leaders of both sides of industry, spokesmen of the media and others who seek to influence public opinion are full of references to turning points in history which, subsequently, stubbornly refuse to turn.

By contrast, however, those who see the events of 1973 as a turning point in the world energy situation can surely expect to be justified by events.

For 1973 was the year when the world's leading oil exporters, members of the Organisation of Oil Exporting Countries (OPEC) grasped their opportunity to push up oil prices to hitherto totally unprecedented levels. Importing countries had no option but to accept the new inflated prices. Cheap energy, it was clear, had gone perhaps for ever, to which was added the recognition that a regular and adequate supply from these sources could no longer be assumed in the future.

Although oil is only one of the various forms of energy available to mankind, it has assumed such a dominant role that a major interruption in availability inevitably affects the entire world economy. For this very reason, the sudden jolt to the general complacency, nurtured on ideas of super-abundant energy, may yet prove to be a blessing in disguise.

It has brought home to everyone the realisation – known already to many of those who have been more closely concerned with energy problems – that the world's supplies of energy, in their present forms, are not nearly so plentiful as has been previously supposed. Moreover, what reserves remain will become increasingly expensive to exploit.

One of the forms of energy which will have a major role in the future is coal, because it has by far and away the largest reserves of any fossil fuel. On the other hand it suffers, compared with oil and gas, from generally greater problems of production, handling and usage. The challenge for coal in the future is how to overcome these problems. In this the need for adequate investment and research is vital.

Although my immediate concern must necessarily be with the place of coal in the United Kingdom, I fully recognise that the energy problem is essentially a world question. What we do in the United Kingdom will have effects beyond our borders and, still more so, what happens in the rest of the world will affect the lives of all of us in this country.

Now that the United Kingdom is a full member of the European Community we have to consider our situation as a part of the wider European energy scene – a scene, however, in which we can expect to play a powerful part with our very large reserves of fossil fuels – coal, oil and natural gas.

It would be comforting to think that the enlarged Community, by pooling its energy resources, could look towards near self-sufficiency, but this is not so. A considerable proportion of the energy needs of the area will continue to be obtained from outside and, for this reason in particular, I shall keep international considerations prominently in mind throughout the book.

Limits to Energy

There are limits to potential energy exports around the world. It is necessary therefore to take a global view and consider how demand from other energy-hungry parts of the world will affect Britain's and Europe's prospects for importing their future energy requirements. It is also particularly important to consider how these international supply problems will affect the price at which supplies can be obtained, remembering that changes in the price of one form of energy can influence the prospects and availability of others.

There are very few parts of the world which are exempt from these influences, though some have a bigger impact than others. Thus the oil producers of the Middle East and Africa in particular, are exporters of huge quantities of energy while Japan exerts a big influence as a major importer of energy. The United States is one of the main energy producing areas of the world, but its use of energy has grown to such an extent that it is now a major importer, with the prospect of becoming possibly even more import-dependent in the future.

At the moment, the USSR and Eastern Europe are largely self-sufficient in energy and do not compete with the rest of the world for OPEC's oil supplies nor does it seem likely that they will do so to a major extent in the foreseeable future. On the other hand, the USSR, is already an exporter of oil and gas and has the reserves "in the ground" to increase its exports quite considerably, and should it do so this would prove to be a significant factor in the world situation.

China has started small exports of oil to Japan and it is possible that it could become a sizable exporter of energy. There are also the vast areas of the world – the Indian sub-continent, Africa and Latin America – where huge and growing populations could eventually exert a big call on world energy reserves, reduced, maybe, by new energy discoveries in their own unexplored regions.

Where energy is concerned, it is very much a case of one world and it is therefore essential to take a broad international view. Equally, while placing my emphasis on the role of coal, I shall also give due weight to the other forms of energy since all have their part to play and each interacts with the others. In particular, the rate at which any single form of energy is consumed affects demand for the others, especially when the alternatives are more costly or not immediately available.

World resources of coal, oil and natural gas – the fossil fuels – were built up over millions of years in the remote past and are not being renewed, at least to a measurable extent. It follows therefore that the available world reserves are finite, while consumption has risen at a great rate as people all over the world have grown accustomed to a way of life which demands more and more energy.

Oil has illustrated this trend more clearly than any other fuel. In the decade and a half up to 1973 it was sold well below the price of many competing fuels, even allowing for the cost of transport and the payment of royalties and taxes to the producing countries. Everyone assumed that these conditions would continue and that the comforting thought, often expressed a few years ago, that more oil is discovered each year than is consumed, would remain true indefinitely.

In recent years, however, not only have major discoveries tended to be in relatively costly areas, such as Alaska and our own North Sea, but the consensus of expert opinion, even before 1973, was that the quantities of oil which could be extracted at costs covered by the then current world prices would not be capable of meeting the rising demand for very much longer.

Possible substitutes for oil (for example, from oil shales and tar sands) did exist but would be much more expensive than oil had been up to that time and, in any case, the time scale for their exploitation was so long that they would not be available in substantial quantity for many years ahead. As matters stood, therefore, given the uneven distribution of oil reserves throughout the world and the high cost of substitutes, oil prices would inevitably have risen over the years. Had not the OPEC countries exploited their monopoly situation as they did, however, the rise in prices would have been more gradual, encouraging the introduction of substitutes, as part of a natural progression which made them economic as oil prices rose.

End of an Era

This possibility ended in October 1973 and the public which had assumed that adequate supplies of fuel at reasonable prices would always be available, apart from temporary local difficulties due to strikes and other emergencies, found that a large proportion of the energy supplies of the UK and many other industrialised countries was dependent on suppliers whose interests and policies could run counter to those of their customers, resulting in steep price rises and even a cessation of supplies.

We are still probably too close to these events to base our future actions on the lessons to be learned from them alone. Thus, to put the world's energy problems in true perspective, I believe we should look both backwards and forwards to encompass the fifty years of the second half of this century.

I see ourselves standing at the mid-point of this fifty-year span from which we can look back over twenty-five years to the start of the post Second World War period and forward to the end of the present century.

This is not just a neat way of dividing up an arbitrarily selected period. On the contrary, it offers an opportunity to examine the trends of the comparatively recent past, to absorb the lessons they teach and apply them to the future. Moreover, the past twenty-five years illustrate with particular clarity the trends in the coal industries of Britain and the world, with which this book is principally, though not exclusively, concerned. The next twenty years on the energy front will be significantly affected by decisions taken now in respect of all forms of energy but especially those, like coal, for which the time scale of development is long.

Looking more closely at our fifty year period, I further sub-divide it into a series of shorter periods. The first, running from the end of World War Two till the second half of the nineteen fifites, was a time of energy shortage, occasioned by the war and its aftermath. This was followed by a period of surplus energy which lasted until the early 1970's when a tight supply situation began to emerge, marked by oil price rises in 1970 and 1971 which, however, were modest compared with what was to come.

With the Middle East War of 1973 came the start of the third of the four periods, confronting the world with an energy crisis, the like of which it had not experienced before. It was this crisis, which ushered in the period through which we are still passing and which, although the immediate effects have worn off, has left the world questioning the whole basis of its energy supplies and viewing the future with justifiable concern.

Nevertheless, in spite of an apparent air of alarm, this can be described as a "contrived" crisis, brought about by political and other pressures rather than by the logic of the present world energy situation. In other words, the world is not short of energy at the present time but is suffering from one of the symptoms of shortage, in the form of high energy prices, because of man-made policies.

What I am saying, is that sooner or later the crunch will come and the world will enter the fourth, and final period of the fifty-year span with which this book is concerned. This I describe as the "real" crisis, when energy shortages will become acute, even if there is no artificial stimulation, unless drastic and far-reaching steps are taken sufficiently in advance.

"Real" Crisis

As the story unfolds, it will be seen that the "real" crisis will become apparent probably in the late 1980's or 1990's and certainly by the opening of the next century. It is impossible to be more precise but I shall discuss the factors which will affect the timing of the crisis in later chapters. Whilst the way in which these factors develop will determine the advent of the serious energy problems, even on the most favourable assumptions, they cannot be long delayed. I believe that the timing will be such that decisions necessary to meet the crisis should be at least under active consideration now while some of the longer term decisions should be taken as soon as possible.

I am thinking in world terms, and, obviously, the onset of the "real" crisis will vary from one individual country to another. It is likely, for example, that Britain's own crisis may be delayed, relative to the rest of Europe, because we shall be self-sufficient in oil and gas, as long as North Sea supplies last, and, of course in coal for very much longer. On the other hand, oil and gas (but not coal) are likely to be on the decline by the end of the century, so that we may find ourselves having to resume imports just at the time when the world's energy situation is becoming acute.

Coal, on this showing, will have a continuing key role as the world's largest single fossil fuel resource and many of the decisions which must now be taken relate to the world's coal industries.

Although the decisions cover a very wide field, from policies for conserving present forms of energy to the exploitation of completely new types, some of the most urgent are concerned with an expansion in the output of existing fuels, particularly coal, and having secured the increased output, making certain that the various fuels are put to optimum use. This will call for an intensive research effort, covering both the production and the utilisation of the energy forms and I shall have much to say on this point.

Research along these lines has been carried out for many years and has been stepped up to a certain extent as a result and important decisions in this regard should be taken now. Further research is also needed on the many problems surrounding the future development of nuclear power, and, although so-called benign and renewable sources, such as solar power and wind and wave power, seem unlikely to make a major contribution to the overall reserves before the end of the century, research into these sources is to be encouraged.

Many of the decisions which the world will have to take in order to meet its future energy needs will affect the environment, at a time when people everywhere are becoming increasingly conscious of the damage which can be done to natural surroundings in the search for and exploitation of the different forms of energy, whether an opencast site, electricity pylons, a natural gas pipeline compressor or a nuclear power station. There is no way of providing the increased energy supplies necessary without some effects on the environment but, as I shall aim to show, the fuel industries believe it is possible to strike a reasonable balance between mankind's need for adequate energy supplies and its justifiable concern for the environment in which we all live. This is certainly our view in the British coal industry.

Environmental protection is only one of the many facets of the world's energy problems which will demand the combined efforts of governments, the energy industries and consumers, acting on both national and international levels, if satisfactory solutions are to be found. The solutions are unlikely to be so obvious and so clearly to everyone's advantage that the necessary co-operation will be easily achieved. On the contrary, conflicts of interest between different countries and between groups within individual countries are certain to arise and will call for the highest qualities of statesmanship and leadership from all concerned.

I believe that the problems can and will be solved, provided they are considered as dispassionately as possible in full knowledge of the facts and that a rational long term strategy is evolved. It is with the aim of making a contribution to public discussion of the world's present and future energy problems that this book has been written. It goes to print in December 1977 and details will inevitably become out of date as events unfold. The broad picture given by the book, with its central theme of the need to press on with greater production of coal and other indigenous fuels in the main energy consuming countries and to keep down demands on the world's dwindling stocks of oil and natural gas, should however remain valid for years to come.

2 Up to 1973

All the evidence suggests that the second half of the Twentieth Century will prove to have been the most vital fifty years so far in the world's energy development. It is my belief that what remains of this fifty year period will be particularly significant for the coal industries of Britain and the world, and will involve a total reappraisal of the role coal is to play in the long term future.

In order the better to foresee this role I propose to trace the story of the past 25 years or so, examining the changing patterns in the world energy scene and, in particular, the fluctuating fortunes of coal during this period. This brief historical survey should help to explain how the world has arrived at its present energy situation and, more especially, provide the springboard for a discussion of the steps which must be taken now to deal with the "real" crisis which, as I explained in the previous chapter, I expect to come upon us, probably during the 1990's.

I look first, then, at the ten years or so after the close of the Second World War which saw the world gradually recovering from hostilities on a scale which could not fail to have adverse effects on the energy economies of every country involved and, indeed, on those few countries which were not directly concerned.

Understandably, the most severe immediate post-war energy problems were found in Western Europe. Not only had the efforts of the coal and other energy industries been directed solely to exploiting whatever resources were to hand, without considering the future, but, on the Continent, and to a lesser extent in Britain, output had been interrupted by direct military action.

For some years after the end of the war, so far as Britain and the rest of Europe were concerned, it was a question of seeking to restore energy supplies to something approaching normality as rapidly as possible, in order to meet the overall energy shortage. Britain's energy policy, like those of the rest of Europe and other countries around the world, was directed towards increasing the output of coal, then still the most important fuel for most purposes, and to encourage economy in its use.

Although the world was no longer as dependent on coal as it had been before the First World War, or even in 1925, when it represented about five-sixths of all commercial forms of energy used in the world, it was still far and away the major fuel when peace came in 1945. As late as 1950, coal accounted for about 60 per cent of total world energy consumption.

In order to make this comparison between different forms of energy, we convert the various fuels into "tons of coal equivalent" (or tce). On this basis, total world energy consumption in 1950 was 2,562 millions of tons of coal equivalent (mtce), of which coal represented 1,544 mtce, or 60 per cent as said above. This basis for comparison will be widely used throughout the book.

Perhaps I should also add here that I shall try to keep my use of statistics to the minimum, consistent with the need to substantiate my views. I shall use United Nations figures wherever I can, and certainly throughout this chapter, as these are available on a comparable basis for all countries from 1950 onwards. Most of the UN figures go only up to 1974, though a few series go on to 1975. Where later figures are given in subsequent chapters, they have been taken from other sources and adjusted if necessary to give a provisional figure on the UN basis.

To return to the story of the past 25 years. After the war strenuous efforts were made to increase coal output, especially in Britain, where the industry was nationalised on January 1st, 1947. The National Coal Board vigorously set about integrating the management of nearly 900 separate colliery companies, all at different stages of development and with widely differing profitability ratings. Between them, they employed approximately 750,000 people, making the National Coal Board (NCB) by far the largest single industrial employer in the United Kingdom at that time.

With its statutory monopoly over coal production in the United Kingdom, the NCB made good progress with the rationalisation of Britain's basic fuel in the immediate post-war years, doing its best to overcome the damaging effects of inadequate investment during the war years and, indeed, for some considerable time before. Nor was the necessary capital available in the amounts really required during the late 1940's and early 1950's, when the need to reconstruct much of British industry placed big competing demands on available resources.

Although I would not like to claim that the coal industry was unfairly treated in the matter of capital availability, it took some time for the momentum of investment to be established.

In view of these difficulties, it is the more surprising, and, I believe, to its credit, that the newly-constituted nationalised coal industry was able to satisfy so much of the nation's demand for energy during the ten years after the war and, for most of the period of energy shortage, Britain's mines were producing well over 200 million tons annually (227 million tons in 1957).

At that time, on a world-wide basis, coal, at nearly 2,000 mtce, still accounted for well over half total world energy consumption of some 3,700 mtce. Nevertheless, it was becoming clear that the mainly coal-based fuel economy of Britain and most of the world was about to undergo a fundamental change during the second of the four periods into which I have divided the 1946–2000 energy scene.

Energy in Abundance

During this period, roughly from the late 1950's to the early 1970's, the world passed through an era of surplus energy even if, in retrospect, the abundance clearly could not last and resources which should have been husbanded for the future were being frittered away.

This period was characterised by a huge upsurge in the demand for energy everywhere and by a considerable, almost dramatic, change in the "mix" of fuels within the total. (see Table 1).

Table 1: World Energy Consumption (mtce)

	1950	1957	1964	1972
Solid Fuels	1544	1926	2223	2406
Liquid Fuels	732	1223	1966	3574
Natural gas	244	447	828	1555
Hydro-electricity	42	68	103	159
Nuclear electricity	--	--	2	18
Total	2562	3664	5122	7712

Source: United Nations *World Energy Supplies 1950–1974* (Statistical Papers Series J No. 19).
The figures include bunkers, but exclude non-energy uses of oil.

Between 1950 and 1972, the world's consumption of primary fuels (these are coal, oil, natural gas, hydro-electric power and nuclear electricity) rose from 2,562 mtce to 7,712 mtce, three times as much. Within this total, however, the consumption of oil rose five-fold, to represent nearly half the total primary energy use by 1972. During the same period, the world's consumption of natural gas increased more than six times, but since it rose from a smaller base, it was still only used on less than half the scale of oil in 1972. The world's hydro and nuclear electricity usage, though, taken together, increasing by four times during the period 1950–1972, still accounted for under 2½ per cent of the world's energy needs, while nine-tenths of the combined total consisted of hydro power.*

Having said this, however, I must point out that world coal consumption continued to rise during this period. From a total of 1,926 mtce of all solid fuels in 1957, consumption rose to 2,223 mtce in 1964 and 2,406 mtce by 1972. The rise in coal consumption, therefore, did not keep pace with increasing demand for energy as a whole so that, from holding a 53 per cent share of total world primary energy consumption in 1957, coal declined to 43 per cent in 1964 and 31 per cent in 1972.

Table 2: Solid Fuel Production in Selected Countries (*mtce*)

	1950	1957	1964	1972
World Total	1580	1951	2239	2439
of which:				
United States	506	468	456	540
USSR	220	376	425	481
United Kingdom	219	228	198	121[x]
West Germany	150	181	177	136
Poland	79	96	123	162
France	52	58	55	32
East Germany	44	67	79	75
China	43	131	290	401
Japan	39	52	51	28
Czechoslovakia	35	53	74	79
India	33	44	63	77
South Africa	26	35	45	59
Australia	19	23	33	62

Source: United Nations *World Energy Supplies 1950–1974*
[x]1972 was affected by a miners strike in the UK.

Declining solid fuel production during this period was unevenly spread around the world, particularly from the middle 1960's when Britain's coal output had been as high as 198 million tons as late as 1964, compared with 131 million tons in 1973. West Germany's production of solid fuel at this time had been 177 mtce (1972 output 136 mtce) and France 55 mtce (32 mtce in 1972).

Outside Western Europe, however, although the consumption of coal fell during the 1950's in the United States, output picked up around

*Most authorities, other than the United Nations Statistical Office convert hydro and nuclear electricity to coal equivalent by a different method which makes them about 7 per cent – a better measure of their importance.

1960 and has been on a rising trend since then. Solid fuel has remained the predominant fuel throughout the post-Second World War period in China, the USSR and in Eastern Europe, particularly Poland, Czechoslovakia and East Germany.

This only serves to throw into relief the extent to which we in Britain and Western Europe neglected our coal reserves in this period, compared with other major coal producing countries. (see Table 2).

Economic considerations explain the divergence between coal output figures in Western Europe and the United States, where coal has generally been cheaper to mine, mainly because coal recovery techniques are largely based on strip mining methods (known as open cast in Britain), which is much less costly than deep mining. The failure of coal output to rise in Western Europe, and Britain in particular, was thus not due to lack of demand for energy in general nor to shortages of coal reserves (in Britain at least).

Europe's coal industries gradually lost their impetus during this period because, generally speaking, the market for coal was not there, within the price situation at the time. A run-down of coal output was therefore inevitable, in the absence of any positive steps by governments, through fiscal means or otherwise, to encourage a greater use of coal in comparison with other sources of energy.

It is not for me to apportion blame, if there is any blame, for what can now been seen as the short-sighted acquiescence in the run-down of Europe's coal industries. The plain fact is that because coal is an extractive industry, continual investment is needed merely to offset the normal exhaustion of mining capacity year by year in order to maintain output at the current level. Still further investment is needed to re-expand capacity and even then there is inevitably a considerable time lag between making the investment and reaping the benefits.

If we learn our lesson from this period of coal's decline something will have been achieved. The need for investment in new capacity today is the greater because of the lack of it in the past.

Oil Takes Over

Of the alternative fuels available, it was oil which reaped the main advantages from the expansion of demand for energy during the late fifties, sixties and early seventies, due largely to its low price and to its near indispensable role in transport by land, sea and air, which had been the real force behind the rise of the international oil industry before the Second World War.

However, oil had not made very big inroads into the bulk heating market, such as power generation and large-scale industrial and commercial heating before the War, because of the relatively high cost of transporting it to Europe, and the quantities in which it was available.

All this changed rapidly after the Second World War, with the rapid development of Middle East supplies, the great expansion of refining capacity in the importing countries and the encouragement given to oil consumption in most industrialised countries. The result was that oil soon began to make significant inroads into those commercial and industrial markets which had hitherto been the domain of coal.

It is only necessary to look at the increase in crude oil production by the leading exporting countries (the Arab States, Iran, Venezuela, Nigeria and Indonesia) to realise the extent of this change in the world energy market. From a total crude production of 170 million tons in 1950, these countries advanced to 473 million tons in 1961 and 1,375 million tons in 1972.

Most of this oil was very cheap to produce and even after royalties and taxation, at the rates current in the 1960's, had been paid to the governments of the producing countries and transport costs added, it could be delivered to consumers in Britain, Europe and elsewhere in the world at prices below the cost of most other fuels, including coal in many applications.

Moreover, most of the countries where the vast new oil reserves were being exploited were under-developed at that time, with low living standards and, having little use for the oil themselves, vied with one another to develop their oil reserves and so secure the revenue from royalties. The result was a glut of oil at low prices.

In Britain, as the period advanced, coal output declined as oil increased its share, though it put up a good fight and as late as 1970 Britain's consumption of primary fuel included 154.6 million tons of coal and 132.4 mtce of petroleum. Only after then did oil move ahead of coal so that by 1973 the comparable figures were 130.4 mtce of solid fuel and 148.6 mtce of petroleum.

It is not generally realised how recent was the deposition of coal from the position of major energy supplier in Britain and this fact deserves to be remembered. (see Table 3).

Other Fuels

Although oil was undoubtedly the major factor in the changing energy situation during the second of the four periods into which I have divided the last half of the twentieth century, other fuels were also making their mark, and contributing to the apparent surplus of energy which was a feature of this period.

Natural gas, for example, of which world consumption in 1925 had been 45 mtce, had risen to 244 mtce by 1950 and to 1,555 mtce by 1972, a more than six-fold increase between the two latter dates. As recently as 1950, however, 90 per cent of the world's natural gas was consumed in the United States and even in 1972 the US still accounted for more than half the total world production.

Elsewhere, natural gas had been discovered in small but usable quantities in Italy, Germany and France during the 1950's but it only began to assume significant proportions in Western Europe with the opening up of the Dutch Slochteren field in 1959, one of the largest discovered anywhere in the world up to that time. The development of this field proceeded steadily and by 1970 Holland had become the world's fourth largest producer of natural gas, after the United States, the USSR and Canada, and has since overtaken Canada to become the third largest producer.

Table 3: Energy Consumption in the United Kingdom (mtce)

	1950	1957	1964	1972
Solid fuels	205	215	188	123[x]
Liquid fuels	25	42	88	145
Natural gas	---	---	---	36
Hydro-electricity	---	---	½	½
Nuclear electricity	---	---	1½	3½
Total	230	257	278	308

[x]Affected by miners strike
Source: *United Nations World Energy Supplies 1950–1974.*

Not the least of the results of this discovery was the impetus it gave to North Sea exploration, once agreement had been reached on the international ownership of the continental shelf, with the right to grant licences for exploration offshore. Britain was particularly active from the start, having an extensive and old-established manufactured (or "town") gas industry, originally based on coal but by this time in the process of changing over to oil-based feedstocks. It had also just become interested in the higher calorific value natural gas through imports of Algerian natural gas in liquified form.

British licences for North Sea exploration were issued in 1964 and, within a year, BP had discovered the first North Sea gas in commercial quantities on the West Sole Field, east of the Humber estuary.

On the strength of this discovery, and the finding later of the Leman Bank, Indefatigable, Hewett and Viking Fields (in which the National Coal Board participated), the then Gas Council (now the British Gas Corporation) put in hand the nation-wide conversion programme, involving the change over from manufactured to natural gas firing of an estimated 40 million gas-burning appliances of all kinds, owned by about 13½ million customers.

This massive process, now completed, was well advanced by 1972 when the contribution of natural gas to the total UK primary fuel input was 36.4 mtce.

To complete the story of the changing energy situation during the period 1957-72, reference should be made to the other major primary energy sources – hydro and nuclear electricity. Hydro power increased from 68 mtce to 159 mtce a year during the period but did not grow appreciably as a proportion of the whole and only represented 2 per cent of the world's energy supplies (but see the footnote on page 10). In Britain it has an almost negligible role, apart from the North of Scotland.

Nuclear power, in world terms, was much smaller still. In 1972 it only accounted for 18 mtce, or about 0.2 per cent of the total, although it has since been growing fast in a number of countries. Everyone will remember however, the heady days when nuclear power was seen as the main answer to the world's future energy problems and this was undoubtedly a factor in the run down of coal.

Since then, there has been a major shift in the whole nuclear power situation, not least in the UK which was first in the field in the commercial exploitation of nuclear power for the public electricity supply system, but this raises big questions which I shall discuss later.

Although hydro and nuclear power, with natural gas, played their parts in the relative decline of coal which characterised the period of energy surplus which ended in 1973, it was oil which had been the dominant factor. It was also to prove to be the central factor in the third period which opened in the autumn of 1973 and in which we are still living.

The "Contrived Crisis"

With the outbreak of hostilities between Israel and the Arab world in October 1973, the energy situation entered an entirely new phase which I described in the preceding chapter as a period of "contrived" crisis. By this I mean, as I briefly explained, that the energy shortages which emerged almost immediately were the result of action by the oil producing and exporting states and did not arise from any sudden deterioration in the world energy supply.

Before considering the actual crisis and its effects, I feel it would be desirable at this point to trace the events which preceded the hostilities and which, though not directly related to the Arab-Israeli conflict, have become an integral part of the situation which followed it.

Looking back, with hindsight, it is clear that although the energy crisis seems to have come suddenly upon the world, in fact, it was inevitable once the industrialised countries, in particular, had allowed their energy supplies to be increasingly tied to oil, with its essential role in so many vital domestic, commercial and industrial activities.

Not only did the world, during the two decades prior to 1973, soak up more and more of the world's oil each year but the main non-oil producing countries, including Britain and Western Europe and Japan in particular, took their supplies from a relatively small number of exporting countries, located, with few exceptions, in the Middle East and Africa (see Table 4). In 1972, the Arab states, Iran and Nigeria produced 1,100 million tons of crude oil, most of which was exported. The only other sizeable exporters were Venezuela and Indonesia which respectively produced 171 million and 54 million tons a year.

Table 4: Production of crude oil by principal exporting countries (million tons)

	1950	1961	1972
Saudi Arabia	27	73	302
Iran	32	59	253
Venezuela	78	153	171
Kuwait	17	87	167
Libya	—	1	108
Nigeria	—	2	91
Iraq	7	49	71
United Arab Emirates	—	—	58
Indonesia	7	21	54
Algeria	—	16	50
Total of countries shown	168	461	1325

Source: United Nations *World Energy Supplies 1950–1974.*

As a graphic illustration of the growing dependence of the world on Middle East and African states, in 1950, Venezuela, with an output of 78 million tons a year had been by far the most important exporter. At that time, the Middle East and African countries had produced only 83 million tons between them and Indonesia only 7 million tons.

By 1961, output in these Middle East and African countries had almost reached 300 million tons – three-and-a-half times the 1950 level. The rate of increase of output was a little higher still from 1961 to 1972 and in tonnage terms the increase was dramatic. Meanwhile, Venezuelan output rose to a peak of 196 million tons in 1970 and then began to decline while Indonesian production built up progressively between 1950 and 1972.

Among the Middle East and African states, Saudi Arabia had become the leading producer, with 302 million tons in 1972, followed by Iran (253 million tons), Kuwait (167 million tons), Libya (108 million tons), Nigeria (91 million tons), Iraq (71 million tons), United Arab Emirates (58 million tons) and Algeria (50 million tons). Apart from Libya, which had reached a peak of 160 million tons in 1970, and Iraq, whose output had stood at 84 million tons in 1971, all these states reported record crude outputs in 1972.

In 1972, only three other countries in the world produced as much as 50 million tons, making little impact on the world wide trade situation in oil. The United States produced 525 million tons but, even so, was a net importer, USSR produced 400 million tons of which about a quarter was exported, but most of it to Eastern Europe while Canada produced 86 million tons, with exports and imports almost balancing each other.

Import Imbalance

Not only was the world's exportable oil concentrated in a relatively localised area in 1972, but the major importers were also limited in number and, apart from the United States, were almost entirely dependent on these sources.

In 1972, the biggest single net oil importing areas were the then six-member European Economic Community (West Germany, France, Italy, Belgium, Netherlands and Luxembourg) with imports of 419 million tons, followed by Japan (234 million tons), the United States (228 million tons) and the non-EEC member countries of Western Europe (150 million tons).

United Kingdom net oil imports in that year – not included in any of the above totals – amounted to 106 million tons.

So the position had been reached where a relatively small number of industrially developed (and therefore extremely oil-dependent) countries, all of whose economies and, to a not inconsiderable extent, their wider policies, were linked, found themselves dependent on another relatively small number of oil exporting countries, which, with some exceptions, might likewise be expected to work closely together in the economic and political sense.

It is easy to realise now how vulnerable the Western world, in particular, had become. There were those, however who, like Lord Robens, (my predecessor as Chairman), E.F. Schumacher (Economic Adviser to the Board 1950 to 1970 who subsequently gained an international reputation with his work on intermediate technology and his book 'Small is Beautiful') and the leaders of the mining unions in Britain had been warning of just this very danger for some years before.

As long ago as 1956, the first Suez emergency had dislocated the movement of oil from those all-important Middle East oil fields to

Europe. Restrictions on the use of petroleum products were introduced and had the effect of inducing some temporary doubts on the wisdom of relying so heavily on an imported form of energy, and more or less from one source into the bargain.

In the event, however, the extreme difficulties which were forecast as a result of the initial shock of the Suez Canal closure never actually emerged with the seriousness expected, largely because the United States and other countries outside the Middle East oil exporting areas were able, at that time, to step up production and arrange for additional deliveries to the oil importers at short notice.

As a result, it appeared to the peoples of Britain and Europe that the international oil companies were well able to surmount any crisis, even of this magnitude, with only temporary restrictions on consumption and with price rises which, though sharp enough at the time, soon fell back to pre-crisis levels, to a point where, once again, oil was cheaper than coal in many applications.

Nevertheless, the lesson of Suez was not entirely lost on the British Government which decided to step up the nuclear power programme which had been instituted on a small scale just prior to these events. Why, it may well be asked, was not some of this investment channelled into the coal industry, particularly in Britain, where it would not only have helped to provide a long term energy supply from an indigenous source as an insurance against outside political interruptions but would also have relieved some of the strain imposed on the balance of payments by rising oil imports?

Succeeding governments, even those which were keen to help the coal industry, concluded, in the light of events at the time, that the decline in the coal industry was inevitable and that the degree of protection, of one kind or another, which would be required to reverse the trend, could not be justified. In the context of the 1950's and 1960's, the risks of interruptions of supply were considered small in comparison with the gains from continuing to use more and more of the then relatively cheap oil.

During these years, oil continued to enjoy its price advantage over coal and, if anything, the advantage tended to increase. Should there be any more interruptions in supply, everyone was sure they would be as short-lived as the Suez example and the oil companies had demonstrated their skill in overcoming temporary crises. Though the Suez Canal was closed again in 1967 as a result of the renewed Arab-Israeli war and did not reopen for several years, its closure had little immediate effect on oil supplies. Governments were also confident that the expanding demand for oil could be met until after the end of the century – which seemed a long time ahead, seen from the 1950's and 1960's.

In any event, oil was essential for petrochemical feedstocks and for transport, and if Britain alone decided to limit the use of oil as a bulk heating medium, it could make no difference to the world supply position but add considerably to Britain's industrial costs, compared with those of our competitors.

Coal, it was thought at the time, could only relieve the situation to a limited extent, in the then state of technological and economic factors. As the years passed, the introduction of nuclear power and, later still, the discovery of North Sea gas and oil, all suggested that the longer term future could be left to look after itself.

It is probable, too, that governments felt that should it be necessary, the decline in coal production could be halted and even reversed without too much difficulty or delay. There are however very good reasons why a declining coal industry cannot be quickly restored to a rising trend, as many were saying at the time.

The important point to stress is that, for whatever reasons, the necessary investment in coal was not made in the UK, nor in many other coal-producing countries, especially in Western Europe, during the 1950's and 1960's.

The Rise of OPEC

I must now make another excursion into history as we build up the story of that period. Just as the oil-importing countries failed to act on the warning that had been implicit in the 1956 Suez incident, neither did they take sufficient heed of the establishment in 1960, of the Organisation of Petroleum Exporting Countries which, under its short title of OPEC, has become only too familiar to energy-hungry countries in more recent times.

It arose out of realisation by the oil exporting countries, based on their experience after Suez, that competition between them for the available international business left each one individually in a weak bargaining position. The present membership comprises, Algeria, Ecuador, Gabon, Indonesia, Iran, Iraq, Kuwait, Libya, Nigeria, Qatar, Saudi Arabia, United Arab Emirates and Venezuela. Between them, in 1972, they were responsible for 43 per cent of total world crude oil production (over 50 per cent of the non-Communist world's output) and over 95 per cent of the world trade in crude.

For some time after its foundation, OPEC made only a limited impact on the world energy scene, though it should have been obvious that its members clearly had the means at their disposal to achieve their declared aims of co-ordinating and unifying ". . . the petroleum policies of member countries and determining the best means for safe-guarding their interests, individually and collectively, as soon as an opportunity should occur."

It was, however, to be another ten years before the chance came to demonstrate their collective power. In the meantime, OPEC concentrated on stabilising the "posted" price of oil. The "posted" price was a notional reference price which was used as the basis for calculating royalties and taxes due to the "host" governments from their customers, the big international oil companies.

By raising the posted price, therefore, OPEC triggered off the prices to the end-user with which the consuming world was soon to become only too familiar.

OPEC on the Move

Changes in posted prices between 1960 and 1970 had only a minor impact on the world petroleum situation but the 1970's ushered in a period which sounded the alarm bells throughout the whole of the petroleum-importing world.

In retrospect, it can now be seen that the second closure of the Suez Canal in 1967 had a cumulative effect on tanker availability and this, combined with a rapidly growing demand for oil, resulted in a general tightness in international oil supplies. The "super-tankers", which cannot use the Canal, even when open, had not arrived on the scene in sufficient numbers by 1970 to handle the increased demand for oil movement across the world.

The international oil companies and the oil consuming countries were thus in a weak position to resist OPEC demands, which, as soon became apparent, were to take two forms – pressure for higher revenues and a greater degree of "participation" (that is, a stake by the host country in the operation of oil companies in its territory). There was also the possibility of actual interruptions in the supply of oil, for one reason or another.

In the summer of 1970, for example, the Trans-Arabian pipeline (TAPLINE) was severed in Syria, remaining out of action for nine months. At the same time, the Libyan Government introduced restrictions on oil production, both to conserve supplies and to accelerate agreement to their demands for increased oil revenues. The policy succeeded, and in September 1970, Libyan posted prices rose by about 15 per cent – the first significant advance in the posted prices of crude since 1957.

About the same time, Algeria nationalised the assets of international companies in its territory and increased the posted prices to French companies by nearly 40 per cent. With other countries unilaterally increasing their posted prices, the scene was set for the historically important OPEC meeting at Caracas, Venezuela, in December 1970 which resulted in a demand for a general increase in prices by all member countries, accompanied by a threat that supplies might otherwise be withdrawn.

This was only a start. Further price increases followed the Teheran Agreement between Persian Gulf States and the oil companies in February 1971 and the Tripoli Agreement between Libya and the oil companies in April of the same year. Both agreements raised posted prices and also included provision for a scale of further posted price increases at yearly intervals for the ensuing five years.

Over and above these scheduled price increases, further separate rises were introduced in January 1972 and June 1973, to offset the effects of the devaluation of the dollar. From June 1973 onwards, posted prices were to be revised monthly and adjusted to reflect changes in currency parities.

Then came the October 1973 war and the real onset of the crisis. On the Arab side, though Egypt and Syria were the principal combatants, many other Arab states provided support and assistance in varying degrees so that the hostilities had widespread effects on oil supplies to the world in general and to Western Europe in particular.

Although the war took place in and around the world's major oil producing region, military action had only marginal effects on the actual production and transmission of oil. Much more damaging was the series of decisions taken both during and after the hostilities by Arab producers in pursuance of their political aims.

In mid-October 1973, oil prices were unilaterally increased by the Arabian Gulf producers by some 70 per cent – a bigger rise than the total of price increases achieved over the previous three years. Libya followed with an even greater price rise.

Hardly had the world absorbed the implications of these price rises when the Arab producers announced their decision to use oil as a political weapon, by progressively refusing oil supplies to countries deemed to be unsympathetic to the Arab cause. This policy was generally directed against Western countries by holding Arab oil production at a figure 25 per cent below the 1973 level, with the imposition of a total embargo on supplies to the United States and the Netherlands.

It is now a matter of history that the combined result of all these measures was to bring about shortages of petroleum products fairly quickly in a number of consuming countries, the precise incidence being dependent on the extent to which each country relied on Arab sources and on their strategic oil stocks at the time. Various measures to economise in the use of oil were introduced throughout Western Europe, including petrol rationing in some countries, though not in the UK, where however, provisional plans were made to the point of issuing coupons.

General cut-backs of Arab oil were relaxed relatively quickly to 15 per cent below the 1973 level but at the same time (January 1974) the most dramatic price increase so far recorded was imposed. Gulf posted prices were more than doubled compared with the previous sharp rises and the Libyan posting, already nearly twice the level of Gulf prices, was also almost doubled again. The oil-importing world had to contemplate a five-fold increase in the landed price of crude oil within a space of some three years.

By July 1974, the embargoes on the United States and Netherlands had been lifted. The world was left with petroleum prices five times higher and an awkward feeling that the interruption in supplies could well be repeated at any time.

Participation

I now turn to the effects of OPEC participation policies which were being applied more or less simultaneously with the tide of rising prices. Perhaps more correctly regarded as a euphemism for nationalisation, participation began when Algeria assumed 51 per cent of all French oil operations within her territory in February 1971, while in August of that year the OPEC countries as a whole claimed 20 per cent participation in all concession-holding companies.

Six months later, in February 1972, the oil companies agreed in principle to 20 per cent participation by host governments, leaving precise terms to be negotiated. The first participation agreement was effectively achieved in January 1973, when Abu Dhabi, Kuwait and Saudi Arabia gained an immediate 25 per cent control of oil operations, with a gradual increase of participation in stages leading to eventual majority control by 1982.

This, in itself, was a notable point in the story of Middle East oil, with a guarantee of majority control over oil production in their territories by a specified date but, as so often in the tide of human affairs, once the initial breakthrough is achieved, a trickle can become a flood.

Even those countries which had been parties to the January 1973 participation agreements began fresh negotiations for immediate 60 per cent participation. Kuwait concluded such a deal in February 1974, Qatar in April and Abu Dhabi in September while Saudi Arabia signed an interim agreement in the same month. Other countries, outside the original signatories, made similar agreements during 1974, including Nigeria, Oman and Bahrain.

Participation, however, did not only mean the partial or complete transfer of ownership and control from the oil companies to the host government. It also involved a further increase in oil prices because the host government sold its share of oil under the participation agreement either back to the oil company concerned or to third parties – in either case at higher prices than those agreed in respect of the share of the oil going to the oil company under the terms of the agreement.

There was thus a distinction between the oil which the oil companies could dispose of as their share in participation agreements (generally known as "equity oil") and the host government's share (described as "participation oil"). Prices of equity oil continued to be based on the notional posted price.

Posted prices have been the basis for the calculation of host government royalties and taxes and these also increased during the crisis. For example, taking Saudi Arabian light crude oil as typical of long-haul Persian Gulf crudes, royalties were increased from 12.5 per cent in June 1970 to 20 per cent in 1974 while taxes rose, over the same period, from 50 per cent to 85 per cent.

Various methods of pricing participation oil (that is, the host government's share) were introduced. Under the original agreements, some participation oil had to be sold back to the oil companies at fixed rates, some at the option of either the companies or the host governments while some was available for sale on the free market. The proportions to be sold under each of these headings varied from one agreement to another and from one date to another, while the prices also varied.

Later, with the reappearance of a crude oil surplus, as a result of the huge price rises, the ending of the initial impact of the crisis and the general economic worldwide recession, the oil companies were obliged to buy back the full 60 per cent of participation oil, or at least a predetermined proportion of it, at prices which, as already explained, were above the tax-paid cost of equity oil.

From 1st January 1975 the OPEC countries introduced a new so-called Single Price structure under which the price differences of participation oil and equity oil were eliminated. The trend towards greater control by host governments has continued, including outright nationalisation (with compensation) in a number of major instances. Profit-sharing contracts, under which oil companies operate on an agency basis, have become more common.

Though these changes have not resulted in major increases in price, as did the earlier ones, they represent a further reduction in security of supply for the importing countries because of the loss of influence of the major international oil companies.

The landed cost of oil in consuming countries includes transport costs and oil company profit margins. Both these factors fluctuate but they have not risen to the extent shown by the cost of crude oil to the oil companies and, with a depressed tanker market, as has happened from time to time, transport prices can even fall.

Where Britain stood

The oil price increases of 1973 and 1974 hit Britain almost as hard as the rest of Western Europe, but we have the area's biggest coal industry and North Sea oil and gas give us the prospect of energy self-sufficiency, at least temporarily, for a few years ahead.

The National Coal Board had already seen the implications of the 1970 and 1971 oil price increases, and had begun to plan for much increased investment in new capacity, aiming first to stabilise and then to increase deep-mined coal output. Britain's "Plan for Coal", referred to frequently in later chapters, gave a lead to the world in facing-up to the energy outlook arising from the high oil prices now and in the future.

As we have noted, North Sea gas was coming ashore in substantial and increasing quantities before 1973. Production of North Sea oil however had not then started (the first oil came ashore from the small Argyll Field in June 1975) although it had been discovered in the UK sector with the Montrose Field in December 1969, and by 1973 it was clear that self-sufficiency in oil during the 1980's was likely.

Aftermath

I have dealt in some detail with the events which precipitated the crisis and have traced its development partly because of the aftermath it has left us and partly because what has happened once could happen again.

In the next chapter I describe what has happened on the energy front since 1973, so completing our background picture of how things now stand.

3 After 1973

In the previous chapter, I carried the post-war story of world energy up to the Middle East crisis of 1973 – the "contrived" crisis, as I have described it. Nothing that has happened since then has caused me to change my view that this event and its repercussions represent a watershed in the history of international energy. Though it is not easy to write about events which in many cases are still unfolding themselves, and where the situation is changing continually even as I write, my aim in this chapter is to trace the main trends on the energy front which have characterised the period since 1973.

For a start, I do not believe that there is a parallel between the post-1974 years and the seemingly comparable periods after the 1956 Suez crisis and the "Six Days War" of 1967, tempting though it may be to look for similarities. Both of the former crises, posed difficulties for the energy importing countries. On each of these two occasions, however, there was a fairly rapid return to normality on the energy front.

It would be easy, today, to suggest that there has, once again, been a similar return to the *status quo*, since there seems to be no actual shortage of energy. Even oil is in adequate supply. In Britain, in particular, it could be thought, despite warning voices to the contrary, that it is once more a case of back to normal, with North Sea oil and gas coming in at accelerating rates and promising self-sufficiency before very long.

The big difference between 1956 and 1967 and the present crisis is that prices have remained high throughout the period, contrary to what many people expected in 1974. In both 1956 and 1967, not only did prices not rise as dramatically during the crisis itself, but they soon returned to a level similar to that which had obtained before. On the contrary however, after the 1973 crisis, with oil prices five times higher than before, there has been no fall back from this unprecedented level, certainly so far as the price of crude oil from OPEC sources is concerned.

The high price of oil cannot be shrugged off as just another aspect of inflation. These higher prices are the result of a fundamental and permanent change due partly to the conscious actions of the major exporting countries, but also partly to the fact, emphasised as recently as April 1977 by President Carter, that world reserves of oil are finite. Even if the "real" crisis is still some time ahead, it is casting its shadow before.

It is understandable that a certain element of complacency should have crept into the thinking of many people because events since 1973 have tended to obscure the true long-term position. The increase in oil prices contributed to, even if it did not entirely cause, the general economic recession which has affected the non-Communist world in varying degree since then. This, in turn, has resulted in a slower growth in demand for energy, compared with the steady and rapid rise of previous years. In many countries demand for energy has actually fallen during the past three to four years.

This can be illustrated by the experience of the members of the European Community (the "Nine") where total energy consumption (not just oil) fell by 2 per cent during 1974, compared with 1973 and by a further nearly 6 per cent in 1975. Imported oil bore the brunt of this decrease in consumption. Western Europe's oil imports fell by 4½ per cent in 1974 and a further 13 per cent in 1975.

Elsewhere in the world, United States oil imports declined by 2 per cent in 1974 and another 1 per cent in 1975. Japan's oil imports in 1974 were down by 3 per cent and by a further 8 per cent in 1975. Taken together, Western Europe, the USA and Japan are the importers who exert the biggest influence on the world energy situation as a whole and, in previous years, declining imports in all these three at the same time would have been expected to lead to a fall in world oil prices. The fact that nothing like this has happened is surely evidence that high oil prices are here to stay, even though there was a temporary surplus of oil and a trend towards slightly lower prices outside those set by OPEC. However, the recession and fall in world demand for oil did act as a warning to the OPEC countries that further massive increases in prices might not be in their own best interests and, compared with previous price rises, the latest increases have been relatively mild.

Nobody should allow themselves to be lulled into a sense of false security. Price rises, if not necessarily at 1973 rates, will be resumed, as demand picks up. As a pointer to the future, United States oil imports reached a new record level in 1976, while the European Community, in spite of Britain's first sizeable supplies from the North Sea, imported oil in quantities well above 1975, though 1976 imports were not back to 1974 levels.

Effects on Policy

During the years which followed the events of 1973, the governments concerned and their indigenous energy industries have realised the danger of relying so heavily on oil imports from a limited number of sources and have also come to recognise that even these sources are themselves not so well supplied with reserves as had been supposed.

If, both at national and international levels, the amount of paper generated on proposals for energy saving had been matched by action, it might not have been necessary to worry about the future. Individual countries have reviewed their energy policies, or more often, considered whether they needed an energy policy at all and, if so, what shape it should take. Regional and international organisations have started to consider policies and to set up machinery for future action, especially in the field of research and development and in the preparation of estimates of supply and demand under varying assumptions.

This is all to the good and even if the practical results have been disappointing so far, at least the world is more aware of the long-term energy problems than would have been the case had there been no "contrived" crisis.

Leading the way in this movement was the United States which initiated Project Independence, the imaginative but, as it soon proved, unattainable objective of reaching virtual self-sufficiency in energy by 1985. It was during the Nixon Presidency that Project Independence was launched but neither he nor President Ford were able to make much headway, against political, economic and environmental obstacles.

It is too early to judge whether President Carter will be able to make a bigger impact on problems where there are no easy options. His early energy policy pronouncements in April 1977, were impressive and immediately aroused keen interest world-wide, some criticism but also a wide degree of agreement.

His first statement concerned nuclear policy. "There is no dilemma today more difficult to resolve than that connected with nuclear power," he said, going on to make it clear that he regarded the problem as one of ensuring that no elements of the nuclear power process will be turned to the provision of atomic weapons. At the same time, he was equally anxious that the world should enjoy the "tangible benefits of nuclear power."

Using the words of the document, the plan is: ". . . to defer indefinitely the commercial reprocessing and recycling of the plutonium produced in the US nuclear power programme; to restructure the US breeder reactor programme to give greater priority to alternative designs of the breeder and to defer the date when breeder reactors would be put to commercial use; to redirect funding of the US nuclear research and development programme to accelerate research into alternative nuclear fuel cycles which do not involve direct access to materials usable in nuclear weapons; to increase US production capacity for enriched uranium to provide adequate and timely supply of nuclear fuel for domestic and foreign needs; to propose the necessary legislative steps to permit the US to offer nuclear fuel supply contracts and guarantee delivery of such nuclear fuel to other countries; to continue to embargo the export of equipment or technology that would permit uranium enrichment and chemical re-processing; and to continue discussions with supplying and recipient countries."

Having pronounced on nuclear policy the President presented the rest of his energy policy a fortnight later. He preceded the announcement to Congress with a television appearance which began: "Tonight I want to have an unpleasant talk with you about a problem unprecedented in our history . . . The energy crisis has not yet overwhelmed us, but it will if we do not act quickly."

These and other sentences from the "fireside chat" show how seriously President Carter saw the energy problem in the United States. "The oil and natural gas we rely on for 75 per cent of our energy are running out . . . . The world now uses about 60 million barrels of oil a day and demand increases each year by about 5 per cent. This means that just to stay even we need the production of a new Texas every year, an Alaskan North Slope every nine months or a new Saudi Arabia every three years . . . . Because we are running out of gas and oil we must change quickly to strict conservation and to the use of coal. . . . Too few of our utilities have switched to coal, our most abundant energy source."

Two days after expressing these views publicly, the President translated his opinions into policies in a speech to Congress. It comprised a series of financial encouragements and discouragments and legislative measures designed to change the pattern of energy usage.

They included taxes to encourage the switch from larger ("gas guzzling") cars to smaller models and a tax on all petrol, related to the rate at which annual consumption falls. For the industrial energy consumer, a combination of taxes and investment credits would encourage a switch from oil and natural gas to coal with tax credits to encourage higher insulation standards and the wider use of solar energy.

President Carter also proposed measures to remove the present artificially low prices for American indigenous oil and gas, held below world prices by law. A three-stage tax and other measures would bring home prices up to world levels by 1980.

In the case of coal, the aim was to raise production by about two thirds to more than a thousand million tons a year by 1985, while at the same time promoting a "major expansion of research" into more efficient ways of using it.

Congress has substantially approved legislation to create a new Department of Energy (DOE) under James Schlesinger. DOE brings together functions of more than 50 federal agencies – in particular the Federal Energy Administration, Federal Power Commission and Energy Research and Development Administration. The House of Representatives has passed the National Energy Bill, giving force to most of the President's proposals, but at the time of writing it has encountered serious difficulties in the Senate.

In discussions I had with leading figures in the energy field during a visit in June 1977 to America I gained the impression that President Carter had succeeded in creating the right climate for national acceptance of most of his energy measures. There are some doubts as to how far the objectives will be achieved in practice. The targets for conservation may be too optimistic, and the objectives for stimulating indigenous production, especially of coal, are complicated by major environmental constraints. Before the summer recess Congress passed an Act to regulate the effects of strip mining – a measure described as "tough, rigid and mean". The Act's environmental advantages are considerable, but there is no doubt it will both add to costs and lengthen the time taken to obtain approval for mining.

Only the future will tell how successful the Carter initiative will prove, both in its effects on the United States and in its international implications. On the international front, the United States had made an earlier impact in the founding of the International Energy Agency (IEA) in the autumn of 1974, which can perhaps be regarded as one of the most significant developments of the post-1974 period. The Organisation of Economic Co-operation and Development (OECD) were asked to take the initiative in setting up the IEA, which was to be open to all OECD member countries.

IEA was established to explore ways in which countries with a common interest in minimising their dependence on imported oil could co-operate to this end. The terms of reference are wide, extending from possible oil-sharing arrangements to programmes of research into better utilisation of existing reserves and the investigation of alternative sources. It has been particularly enterprising in the research field as I shall explain in chapter 8.

While governments and the international and regional bodies have in the meanwhile, discussed policies, prepared studies and even drawn up plans, the hard economic realities of the energy situation have, so far, had more success in inducing energy economies and encouraging the development of indigenous resources. If the unprecedently high energy prices of the years after 1973, now likely to be a permanent feature of the situation, have had this effect already, can we assume that the energy industries of the world will themselves solve the world's problems, without active policies agreed and implemented by governments and international agencies?

Events to date have not been reassuring. Nothing has happened during the past three to four years to change my view that vigorous national and international action is still needed if we are not to be caught unprepared by the "real" energy crisis before the end of the century.

I base this belief on recent developments world-wide in the main fuels which I summarise here, in preparation for more detailed discussions in later chapters.

Oil

On the supply side of the oil equation, the outstanding fact is that there has been no new oil discovery anywhere in the world since 1973 which significantly changes the reserves situation. It is true that the North Sea discovery, announced before the crisis, of course, has proved to be bigger than had been originally believed. Nevertheless, these reserves, welcome though they are, to Britain and the world also, to the extent that they will take the UK off the world market for a while, represent less than 4 per cent of world reserves.

Enough is now known about the North Sea to form a broad view of ultimate reserves, and those seeking a new bonanza are turning their eyes elsewhere, for example, to Mexico where, early in 1977, the Director of the Mexican State oil company, Petroleos Mexicanos, startled the world with the news that probable reserves could be "far superior" to 60 thousand million barrels, or twice as much as the United States combined proved reserves of 31 thousand million barrels at January 1st 1977, as given by the *Oil and Gas Journal*.

However, these are probable reserves and, judging from Mexican reports at the time, considerable uncertainty surrounded these estimates (see chapter 5 for the important distinction between "proven" and "probable"). The point I must make is that even if the highest Mexican hopes are realised, as, for the world's sake one must hope may be the case, the reserves will only add some 10 per cent to the world total of proved reserves.

There have been a number of smaller discoveries, regarded as significant in the context of the economies of the countries where they occur. Reports of this type have come from, among others, India, offshore Latin America and in South East Asian waters and although these can be welcomed they are unlikely to have a major effect on the main oil supply problem, dominated as it is by the import needs of the United States, Western Europe and Japan.

Almost all of these are still speculative as I write this chapter and must be set against the world proven reserves which only rose by about 2½ per cent (by 2.3 thousand million tonnes) between Janaury 1st 1974 and January 1st 1977.

Hopes have been expressed during the four years since 1973 that major additional reserves may yet be proved in the United States, but although exploration has been stepped up, both on and offshore, US oil production and proved reserves have both been declining. High hopes of a major contribution from oil shale in the United States and from tar sands in Canada have not materialised.

I shall be referring to all these potential reserves in more detail in later chapters, including the undiscovered oil reserves believed to exist in the USSR where, however, proved reserves of oil in 1976 were slightly lower than they were three years previously. USSR oil production has increased since 1973 but the level of output has hardly matched the combined demands from the home market and exports to her Eastern European allies and to the West.

Even more problematic is the question of China's oil reserves and exploitation plans. Oil production increased during the years since 1973 and exports to Japan have started though it is too early to say with any certainty how important a role Chinese oil exports to Japan or, indeed, elsewhere in the world will have in the future.

Fears of an impending oil shortage have raised hopes of finding large deposits of oil under the deep oceans, as distinct from the continental shelf reserves now being exploited in the North Sea and elsewhere. Assuming there are such reserves, both the time scale of exploitation and the costs of extraction will put them out of the reckoning until well into the next century.

This is only one of the many unknown factors in the world oil situation as it has developed during the years since 1973. Yet even if the best possible construction is put on these developments, I see no reason to change my view that the "real" crisis will be upon us before the end of the century.

Natural Gas

I draw similar conclusions from a survey of natural gas developments since 1973. It is true that world reserves of natural gas have grown during this period by about 270 trillion cu. ft., or 6.5 thousand million tonnes of oil equivalent, an increase of about 13 per cent.

It is significant that the whole of this increase is reported from two areas – the USSR, with increased reserves of 212 trillion cu. ft., and Iran, with additional 60 trillion cu. ft. The rest of the world shows a series of smaller increases in some areas, counter-balanced by falls in reserves elsewhere.

Interest in the huge USSR natural gas reserves, which represent 40 per cent of total proved world deposits, in addition to the enormous undeveloped resources, as a potential contribution to future world energy supplies has risen during the post-1973 period. USSR exports to Western Europe by pipeline have increased during this period and contracts for further supplies have been agreed for some years ahead, though not on a scale to solve Europe's, let alone the world's future energy problems.

Although the Middle East countries do not have the same near monopolistic hold on the world's natural gas export potential as they have on its oil, their role is still an important one and particularly Iran which, as an OPEC pace-setter in oil policies may be expected to play an important part in the pricing and other aspects of natural gas exports. It seems unlikely that gas will be released to the world market at prices which will under cut oil from the same region.

As in all energy discussions, the role of the United States in natural gas is as vital as in any other fuel. United States proved reserves of natural gas fell during the years after 1973 and production has fallen also. The seriousness of the situation was brought home to the American people during the winter of 1976/77, when the unusually inclement weather drastically affected supplies to all consumers and although this may have been largely due to distribution problems, the experience served to focus attention on future availabilities.

US undiscovered reserves were previously thought to be very large but estimates of these have been revised sharply downwards during recent years. The United States has therefore reached the point during these past few years when she must either cut back on the amount of natural gas used in the domestic energy market or make bigger demands on the outside supplies to which the rest of the world is also looking.

There have been numerous signs of growing US interest in sources of outside supply, including liquefied natural gas deals with North Africa and a much more ambitious project, at present only at the speculative stage, for financial and technical assistance to the USSR to exploit the huge Siberian reserves. The scheme itself is a major undertaking which envisages the piping of gas from Western Siberia to a point close to Murmansk, near the Finnish border, for liquefaction and shipment to the United States.

For natural gas, as with oil, the last four years have brought the recognition that here is an energy resource which, though much less fully developed than oil, is nevertheless present in finite quantities. However, within these limits, there is scope for an increase in world production as long as reserves last. The big problem is that much of the world's potential reserves of gas will be expensive to extract while other reserves will be costly to transport from the areas where they occur to points of use. Yet other reserves will be costly both to extract and transport.

Nuclear Power

This is undoubtedly the energy source which has suffered the biggest disappointments during the years since 1973, when, in the wake of the crisis, many countries adopted ambitious plans to develop nuclear power as the answer to their future energy needs. For one reason or another, the nuclear power capacity planned by the United States, France, Germany, Japan and others is falling behind schedule. The world nuclear capacity planned to come on stream by 1985, even in pre-crisis times, will only be about half achieved by that date.

Several important factors contributing to this disappointing rate of progress have emerged, including lower estimates of electricity growth, the increasingly high costs of nuclear power station design and construction, a number of unexpected technical difficulties and – perhaps potentially the most significant of all – the growing public anxiety over environmental and strategic issues.

To illustrate the decline in nuclear expectations, one can instance the experience of the United States. Whereas in January 1972 a Department of the Interior publication estimated that nuclear power output in 1985 would reach 1,982 thousand million kWh, by the end of 1972 the Department's estimate had already been reduced to 1,130 thousand million kWh.

Then, when the Federal Energy Agency's Project Independence Report was issued in November 1974 – approximately a year after the energy crisis – the target was raised again, with a 1985 "business-as-usual" forecast of 1,251 thousand million kWh. Fifteen months later, the FEA's *1976 National Energy Outlook* reduced the 1985 nuclear forecast (in the "reference case") to 868 thousand million kWh. Their most recent estimate is understood to be lower still.

Reductions in the Japanese nuclear programme have been even more dramatic. Forecasts by the authoritative Institute of Energy Economics in 1972 estimated that nuclear power would produce 453 thousand million kWh in 1985. By 1974, in spite of the stimulus of the oil crisis, the forecast had been reduced to 394 thousand million kWh. Since then, the problem of finding the necessary sites, allied to public opposition which has been as strong in Japan as anywhere, has resulted in a big cut-back in nuclear plans. Forecasts published by the Institute in March 1977 show a drastically reduced estimate for 1985 of only 154 thousand million kWh.

Western European experience is also in line with the general post-1973 story of nuclear power. Many governments had concentrated on nuclear power as the one hope for reducing dependence on imported oil. Ambitious plans were drawn up but the targets set at that time have proved altogether too optimistic. For example, the European Commission proposed in their policy objectives of May 1974 that installed nuclear capacity should reach at least 200 GW by 1985. That was an increase of some 60 per cent over the Commission's pre-1973 forecasts which had themselves appeared unlikely to be attained at that time.

In 1975, projections by EEC member countries showed that the 1985 total would now be less than 160 GW. By October 1976, the Commission's forecast was down to 125 GW and in February 1977 the *World Energy Outlook* of the OECD suggested that the EEC 1985 would be lower still, at rather under 90 GW. This figure represented nuclear plant already operating, under construction or on order in the latter part of 1976. Because of the time scale, some of this capacity might not be fully operational by 1985.

Another factor which has recently emerged has been increasing concern about future supplies of uranium. An OECD Report (*Uranium: Resources, Production and Demand 1975*) expressed considerable doubts on whether uranium production could be expanded fast enough to supply the nuclear stations planned at that time, with their long-term requirements. The setback in world nuclear plans will postpone the date when shortages become serious, but it may still be much too soon for comfort, especially in view of President Carter's proposal virtually to scrap earlier plans for fast breeder reactors. It was the fear of a uranium shortage which prompted the interest in the fast breeder reactor which has since, however, been counteracted by strategic apprehensions.

It has done no harm to remind ourselves of the nuclear targets which were being discussed around the world after the 1973 crisis and to realise how far we have fallen behind them. This may be inevitable, for the reasons I have given, and it may not prove too serious a setback, provided energy demand does not rise too rapidly in future years and provided also that we take the necessary steps to make good the nuclear shortfall by developing other resources.

In particular, it will be necessary to pay greatly increased attention to the world's coal resources – the one really abundant and readily available source of energy. What, then, has been happening to the world's coal during the years since 1973?

Coal

Coal figured prominently in the post-1973 plans of all the major energy-importing areas, starting with the United States where Project Independence, both as originally envisaged and as later amended, looked to increased production of indigenous coal as part of the whole package. President Carter's proposals, as we have seen, look to an increase in coal output of two-thirds by 1985, to reach a total annual output of 1000 million tons. Whether environmental restrictions will enable this to come about is being vigorously debated in the United States.

Although Japan also expects to increase her use of coal, this will only be achieved by an increase in imports since her own coal industry has been contracting rapidly and is expected to produce no more than about 20 million tonnes during the 1980's, or the same level as 1974 output. In its latest forecast, the Japanese Insittute of Energy Economics expects coal imports to be stepped up to reach about 90 million tonnes a year, mainly of coking coal for the big steel industry but also considerable quantities of power station grades. The main sources of coal supply seem likely to remain Australia and USA, as now.

Common Market policy objectives for 1985 also looked in 1974 for a swing of power stations towards coal as part of an overall plan which envisaged a small increase in coal production to about 260 million tonnes a year with net coal imports also increasing to about 55 million tonnes a year by 1985. It now looks as though coal production on the Continent will fall below the target, unless more positive measures are taken.

If we look at the world as a whole, we note that the coal industry has been on the upgrade during the post-1973 period. In 1976, for instance, several countries achieved record outputs. United States output in that year, at about 600 million tonnes, showed a 10 per cent increase on 1973 figures. Other countries which showed big increases on the 1973 performance, though starting from considerably smaller bases than the United States, included India, with about 100 million tonnes in 1976 (30 per cent above 1973), Australia, 75 million tonnes including the coal equivalent of lignite (20 per cent increase over 1973) and South Africa, also about 75 million tonnes (20 per cent up on 1973). The Soviet Union and Eastern Europe, taken together, showed an estimated increase of 5 per cent in 1976, compared with 1973.

On the other hand, the record of Western Europe since 1973 has been less happy. Hard coal production in the Community in 1976 was 8½ per cent down on 1973, with lower outputs recorded in the United Kingdom, West Germany, France and Belgium, while Holland has ceased coal production altogether.

This is undoubtedly a disappointment to those who believe in the valuable contribution which coal can make towards the solution of Europe's – and the World's – energy problems. The current position is heavily influenced by the lack of investment in the 1960's and only serves to underline the vital importance of allocating sufficient resources to the development of coal mining so that the necessary output will be available when the need is greatest towards the end of the century.

Other fuels and energy conservation

In the early years of the "contrived crisis" there was a flood of suggestions from some quarters seeking to show that the dangers could be averted painlessly by some particular form of energy (solar energy for example) not at present being harnessed on a big scale. For the most part these proposals failed to take into account the vast amounts of capital resources that would be needed and the disturbance to the environment that would result. The flood has tended to dry up as more authoritative assessments have been released, in Britain and elsewhere. These show that at best the new forms of energy might make a modest contribution by the end of the century, though the economics are still uncertain.

Hydro-electricity is different. This will continue to expand but will remain a minor element in the World's energy balance.

Energy savings (conservation) can have the same effect as additional energy supplies. Recent studies, such as the IEA's 1976 Review of Energy Conservation, have concluded that energy conservation is likely to be more economic than many of the supply expansion possibilities being considered, and have produced a sizeable list of measures that might be taken country by country. So far however relatively little effective action has been taken by governments worldwide, though President Carter's initiative is a major step in the right direction.

Britain's Four Years

As I explained in the last chapter, Britain, equally with the other Western European countries, allowed her energy economy to become increasingly dependent on imported oil during the years of apparent energy abundance prior to 1973. When the crisis hit us, though we experienced little in the way of actual oil shortage, the five-fold price increase in oil set up a chain reaction from which we were still suffering four years later.

Government and the energy industries recognised the need for action, both to deal with the immediate problems and the longer-term issues of over-reliance on a single source of energy (oil from OPEC countries). At the same time, it was realised that Britain occupied a special position in that North Sea oil would make us independent of OPEC oil for a time, though it was being recognised that it was necessary to start preparations against the day when North Sea reserves would begin to decline.

Action was taken by the Government, both to conserve existing energy supplies and to initiate policies for the better utilisation of indigenous resources. One of the most important moves was to establish the Advisory Council on Energy Conservation (ACEC), which was given the task of recommending to the Secretary of State for Energy an energy conservation programme. This was a complex task at a most difficult time particularly for a body which had advisory but not executive powers. The Council under the enthusiastic guidance of Professor Sir William Hawthorn succeeded in creating a more energy conscious climate in the country and has made a real contribution to energy saving. I am happy to have served as a member of this Council.

Direct conservation measures were few in number, the most important being the reduction of speed limits on specified roads other than motorways (since modified) and the setting of an upper limit of 68°F. for central heating systems in non-domestic buildings. The actual fuel savings achieved by these measures may not have been large but the psychological effects may have been considerable, especially in the earlier stages.

Other action taken included financial encouragement to improve insulation in existing buildings and the raising of insulation standards for new building construction. Savings from these sources are difficult to establish but it is known that many fuel efficiency campaigns were set in hand for individual enterprises and reported savings of 5 to 10% were frequently mentioned.

Measures of this kind were probably the best which could have been expected in a democratic country where the limits of compulsion are soon reached and where the weapons of exhortation and encouragement must be widely used. The promotional "Save it!" campaign, the effects of which, again, cannot be quantified, kept the need for energy economy firmly before the public conscience.

There have been some notable examples of good energy husbandry from Government bodies, such as the Property Services Agency, which looks after the Government estate, from Ministry buildings to Army barracks. The PSA has acted in various ways, from simple control measures such as reducing the energy used by Civil Service tea urns to complex Optimum Start systems to ensure economic operation of central heating in large buildings.

Obviously, the rising cost of energy has itself been a deterrent against energy waste during the past four years, especially in industry, where hard-headed self-interest, not to say commercial survival, have been powerful forces in energy conservation. The higher cost of energy is the most important single factor in stimulating greater efficiency in usage.

Reference should also be made to certain changes in the machinery of government, including the establishment in 1974 of the Department of Energy, a revival of the former Ministry of Power which had disappeared as a separate entity in 1969.

With the Department of Energy providing a focal point for future action, the Government next proceeded to make a number of separate announcements on various aspects of the energy situation, with special reference to offshore oil and gas, nuclear energy and coal.

For example, the 1974 Report on *United Kingdom Oil and Gas Policy* dealt with three main aspects of the problem, namely, policy on Government participation in offshore oil development, changes in Corporation Tax and measures to help Scotland and other development areas. Of these three points, the first two were of greater historical significance since they led to two important 1975 enactments – the Oil Taxation Act and the Petroleum and Submarine Pipelines Act.

Under the Oil Taxation Act, the Petroleum Revenue Tax (PRT) was introduced, designed to secure a reasonable share of offshore oil revenue for the nation while leaving the oil companies a fair return. The Petroleum and Submarine Pipelines Act, coming into force on January 1st 1976, extended the Government's powers to control the exploration for and production of offshore oil and provided for the establishment of the British National Oil Corporation (BNOC). BNOC, which will eventually become a fully integrated national oil company, has acquired the National Coal Board's subsidiary NCB (Exploration) Ltd. and has already started to acquire 51 per cent participation in existing commercial oil fields and will have a similar share of costs and profits in future licences.

There will be more to say on offshore oil policy later but in this chapter on recent history it is worth reminding ourselves that the first North Sea oil came ashore in June 1975, making Britain an oil producer in her own right, with the prospect of self-sufficiency by 1980. The build up of supplies from then on is covered in Chapter 6.

Natural gas, on the other hand, first came ashore from the West Sole Field in 1967 and was thus an established feature of the national energy scene well before the 1973 crisis. Direct Government action on gas has not paralleled that taken in respect of oil. This is because the purchase, marketing and distribution of gas has been in the hands of a single nationalised industry from the start.

Deliveries of offshore gas rose steadily from 1967 to 1974 but since then deliveries have increased at a somewhat lower rate and some constraints were necessary on the offer of large loads to industry, as British Gas awaited big new arrivals from the Frigg Field, now coming ashore, and from Brent, in 1980 or soon afterwards.

Government's thoughts on nuclear power development were set out in July 1974 in a report entitled *Nuclear Reactor Systems for Electricity Generation*. In view of the debate which has ensued and which is still unresolved, the contents of this Report may have an historical interest only. For the record, however, the Report announced that Electricity Boards should adopt the pressure tube Steam Generating Heavy Water Reactor (SGHWR) for their next generation of nuclear power stations, starting with units in the 600-660 MW bracket, rather than larger sizes, in order to reduce the scaling up problems from the successful 100 MW prototype SGHWR, in operation at Winfrith since January 1968.

At this time, the only nuclear power stations actually operating were the 11 Magnox reactor stations which, at the end of 1974, had produced almost 30 per cent of the world's nuclear electricity and, during that year, accounted for 10 per cent of the total electricity generated in England and Wales. Magnox reactors, using natural uranium, represented the first generation power stations. The next generation of advanced gas-cooled reactors (AGR) use enriched uranium and five such stations, of an installed capacity of 6,000 MW (compared with the 5,000 MW-plus of the Magnox stations), had been ordered long before the crisis.

Two AGR stations, at Hinkley Point, Somerset and Hunterston, Ayrshire, started to deliver power to the grid early in 1976, and, as we have seen, the decision on the next generation is still awaited.

Britain's Coal

I feel I can look back with some satisfaction on the action taken with regard to Britain's coal since 1973, though I must confess that I share the disappointment of many, both inside and outside the coal industry at our recent productivity performance. This raises big problems, but they are by no means insoluble.

Our first, and logical step, was an investigation of the future of the coal industry in the form of an Interim Report prepared by a Steering Committee representing the Government, the NCB, the National Union of Mineworkers, the National Association of Colliery Overmen, Deputies and Shotfirers and the British Association of Colliery Management, under the chairmanship of Eric Varley, then Secretary of State for Energy. This collaboration between Government, coal industry management and unions, carried on into subsequent planning discussions, was an early example of industrial participation on a tripartite basis.

An *Interim Report* in June 1974 was followed by the *Final Report* which set out the future of coal in Britain in the light of demand in the new energy situation, discussed progress and future work and the need for research and development.

In particular, it endorsed the NCB's *Plan for Coal*, which management and unions in the coal industry agreed with the Government in mid-1974 as the broad strategy for the development of the industry.

Believing that there is an overwhelming case for the coal industry to play a much larger part in providing long-term national energy supplies, the Plan aims first to stabilise and then to increase the output of deep-mined coal. I stress the importance of stabilising coal output before we can even begin to think of expanding production, because our existing pits are on average some 80 years old, and, because of the lack of sufficient investment in the past, many of our mines are approaching exhaustion. As a result, output losses through exhaustion of capacity are likely to average between one and three million tonnes a year. Unless the decline is arrested, losses through exhaustion could accelerate still further and affect many of our most productive collieries.

To remedy this situation the Plan essentially proposed to introduce up to 40 million tons of new capacity during the ten year period from 1975 to the mid-1980's, partly by extending the life of existing mines and partly by entirely new sinkings. In order to do this a very substantial exploration programme was put in hand. The Plan also envisaged an expansion of open cast production from the 1974 output of 10 million tons to 15 million tons by 1985.

Events move swiftly in the UK coal industry, and whereas our sights were set on 1985 in 1974, when Plan for Coal was announced, it is now clear that we must look still further ahead to 2000 AD. This will require not only the full implementation of Plan for Coal but the commissioning of about 60 million tons of additional new capacity during the fifteen years between 1985 and 2000, if the industry is to achieve long-term expansion.

Though this may seem a long way ahead, by which time much may have happened to the national and international situation, it is by no means too far ahead for the necessary planning to be started and, to this end, we drew up Plan 2000 which we issued in February 1977 as our view of the long term policies for coal which the UK should adopt.

I believe that both *Plan for Coal* (which goes up to 1985) and *Plan 2000* are vital to Britain's future energy policies and I shall return to them later. It is because of their importance, not only to Britain but also to Western Europe and the energy-importing world, that I have included them in this historical survey of the energy picture since the end of World War Two and which this chapter brings to an end.

We now see what has gone before. My next task is to draw up an "energy equation" which relates the world's demand for energy during the rest of this century to potential reserves and the possibilities of exploiting them to best advantage. In the next chapter, I investigate one of the three factors in the equation – the future growth of world energy requirements.

4 The World's Energy Demands

Although supply and demand are intimately inter-related, it is necessary to consider each separately and I am dealing with future demand in this chapter because we need to know whether the likely supplies of energy will match the world's requirements. Moreover, we must form an opinion, however uncertain it may be, about forward demand because of the inevitable time lags which occur when developing new sources of supply.

A study of future demand is also of particular importance in a book which is concerned especially with the effects of world energy prospects on the future of Britain's coal industry which, like every coal industry throughout the world will be affected by trends in demand for energy of all kinds, as the prices of other fuels increase under the influence of rising demand, accompanied by prospective shortages of supply.

In the world picture, price rises are likely to be most dramatic in the case of oil, as has been evident since 1973. As oil becomes more expensive, which on present estimates is likely, the demand for coal will rise, since it should become more competitive with oil and other fuels in short supply, over a wider range of applications, previously reserved for oil. These will include uses in which oil scored on the grounds of relatively low cost and convenience in use and, although the factor of convenience will remain, it could well be outweighed by economic considerations.

Most of the countries outside the Communist sector rely heavily on the OPEC countries for their oil supplies so that any increase in demand from these oil importing states will press on a source of supply which has already been a cause of anxiety. The examination of demand trends must be conducted on a world scale because so many different countries are inextricably involved both on the supply and demand sides. Though oil must be central to all our discussions, it will be necessary also to consider the world's changing requirements for other forms of energy. The investigation must also encompass possible demand changes between the major energy-importing areas, especially the

United States, Western Europe and Japan, whose calls on the world stock will particularly affect the world position.

Though forecasting is an essential feature of planning the future development of any source of energy, there is no denying that such forecasts cannot possibly be precise. Nevertheless, it is possible to produce forecasts which at least give an indication of the direction and likely scale of action which will be required. Then, with regular updating as new information becomes available, our forecasts will provide a necessary basis for forward planning.

As an aid to the forecasting process, we can first consider the main factors which help to determine the rate at which demand will increase or decrease. These include the amount and nature of future economic growth, in which population changes are an important contributory factor, and the success or otherwise of national and international energy conservation measures. Relative price trends will also be a major factor.

Although the effects of economic growth, conservation and price on demand are, in themselves, relatively straightforward, the problem of forecasting the likely extent of these changes is much more difficult, as is the assessment of the inter-relation between them. I can illustrate this by considering the problem of economic growth.

Economic Growth

Past experience indicates that there is a close relationship between economic growth and demand for energy, although the correlation is not exact. For example, changes in Gross Domestic Product (GDP), a good indicator of economic growth, may not be precisely matched by movements in energy demand. This is illustrated in the UK over the past decade when GDP rose by 2.7 per cent per annum, while energy demand rose by 1.9 per cent a year over the same period.

Though the rates of economic growth and energy demand may not coincide, they usually both move in the same direction and the movement of one is a useful pointer to the likely changes in the other.

In the period since the last war it tended to be assumed that, allowing for the temporary ups and downs of booms and recessions, world economic growth would follow a steady upward trend. So, too, would the world's energy consumption.

World economic growth in the fifties and sixties was fostered by the abundance of cheap energy – a temporary phenomenon, as we now know – which characterised those years. Now that the years of surplus energy have receded, it would clearly be unwise to base our extrapolations of future economic growth on pre-1973 figures.

Instead, we must see what lessons can be learned from more recent experience. The general recession through which the world has been passing has been partly due to a cyclical downturn in economic activity and partly to the effects of the five-fold increase in oil prices. How much can be allocated to each of these factors cannot be determined.

It must be assumed that the world will emerge from the cyclical recession, though the recovery may be delayed by the effects of high energy costs and this may explain why promises of coming economic revival have proved false on several occasions during the past year or two. Nevertheless, the recession will end – there are signs of this at the time of writing – and the world as a whole may well have to grow accustomed to higher energy prices and even come to regard these as the norm.

I expect that when we are able to look back from the year 2000, it will be seen that the general trend, when the variations are levelled out, will again take the form of a definite upward line, but at a considerably lower rate of increase than during the preceding post-war decades. This slower rate of progress will be due to higher energy prices as much as to any other single factor. For this reason, I expect growth to slow down appreciably towards the end of the century as oil prices – and other energy prices under the influence of oil – rise still faster and, unless effective action is taken now, as real energy shortages begin to appear.

To think in terms solely of a world economic growth rate, however, is an over-simplification, since there are considerable variations between one region and another and between different types of economy, each with its own rate of growth and therefore future demand for energy. There is no guarantee that existing patterns of economic growth in the world's major regions will continue as in the past, and a change in growth rates in a particular type of region, either upwards or downwards, could profoundly affect our energy demand forecasts.

Nor does all economic growth have the same influence on energy demand. An industrialised economy makes far greater demands on energy than does one which is based on agricultural and other primary products. This statement might seem to be too obvious to mention apart from making the important point that any tendency towards increased industrialisation anywhere in the world will mean an extension of economic growth of the kind which makes big demands on energy. All around the world, there are numerous countries, at present mainly engaged in primary production, whose official national economic plans include provisions for increased industrialisation.

Many of these factors, and others of a like nature which could be quoted, are largely a matter of speculation at present. As significant changes in economic growth rates due to any of these causes become apparent they need to be incorporated in the general forecast of energy demand, on an updating basis.

Over the period 1950–1974 the world's population increased by more than a half – almost 2% a year – while energy consumption per head doubled. Much the same rate of growth in world population is expected to continue for the remainder of the century. Thus even if there should be no further increase at all in energy consumption per head, the total demand for energy by the year 2000 would be some 60% higher than in 1974. And inevitably rising standards of living worldwide will mean higher energy consumption per head, so that total use of energy must grow greatly.

Energy Per Head

Another factor which may have a bearing on future energy demand is the wide disparity between *per capita* consumption of energy in different countries. Let us examine the position, using 1974 figures as the most recently available on a comparable basis.

In that year, in terms of tons of coal equivalent (tce), *per capita* energy consumption for the world as a whole was 2.06 tce, having more than doubled since 1950, when it had been 1.0 tce per head. These figures, however, obscure very wide differences between countries.

Table 5. Examples of Energy Consumption Per Head (*tce*)

	1950	1975
World	1.00	2.06
USA	7.32	11.49
Canada	5.33	9.82
U.K.	4.36	5.46
USSR	1.59	5.25
EEC (Nine)	2.37	4.88
Japan	0.55	3.84
Latin America	0.34	0.85
China	0.08	0.65
India	0.10	0.20
Pakistan)	0.03	0.19
Bangladesh)		0.03

Source: UN Series J No. 19 Table 2.

At the head of this league, as of so many others, stands the United States (11.49 tce in 1974), followed by Canada (9.82 tce per head), both far ahead of any other major country.

At the other extreme, come the Third World countries, such as India (0.20 tce per head), Pakistan (0.19 tce) and Bangladesh (0.03 tce). European countries occupy a middle position, typified by Britain (5.46 tce in 1974). The EEC as a whole (i.e. the present nine members including the UK) had a 1974 *per capita* consumption of 4.88 tce. The USSR's figure was 5.25 tce per head and Japan 3.84 tce.

These figures relate to a particular year and for forecasting purposes we need to know how they are likely to change in the future. Past experience may not be the best guide to the future but it could be a contributory factor.

Comparing changes in *per capita* consumption between 1950 and 1974, United States energy consumption rose from 7.32 tce per head to 11.49 tce, Canada's from 5.33 to 9.82 tce and the EEC from 2.37 tce per head in 1950 to 4.88 tce in 1974. Starting from a low base, Japan's energy consumption per head rose from 0.55 tce in 1950 to 3.84 tce in 1974 and, over the same period, USSR *per capita* energy consumption moved from 1.59 tce to 5.25 tce.

Elsewhere in the world there are countries with very low *per capita* energy usage, but large and in some cases rapidly growing populations. In China, for instance, consumption per head has risen from 0.08 tce in 1950 to 0.65 tce in 1974 while in India the corresponding figures are 0.10 tce to 0.20 tce. Latin America (excluding the Caribbean area) showed a rise from 0.34 tce in 1950 to 0.85 tce in 1974.

If these and other countries with present low energy consumption per head record only a comparatively small increase in *per capita* consumption between now and 2000 it would have a significant effect on the world energy situation while an increase on the scale of the USSR or Japan over the past twenty years could be dramatic in its consequences.

Industrialisation programmes in these countries will call for more energy to power the new machines and also encourage urbanisation, with a corresponding rising demand for energy to sustain the services an urban population needs. At the same time, transport systems and other infra-structure facilities are required, all of which require energy for their construction and subsequent operation.

With growing affluence, other energy-consuming factors emerge, ranging from wider car ownership and greater use of road, rail, sea and air transport to higher standards in the home, such as better labour-saving appliances and central heating, where this is climatically necessary. All the countries with very low *per capita* consumptions have still to experience these energy-intensive trends.

When it is realised that the countries about which we are talking probably account for about 60 per cent of the world's population, the additional demand for energy brought about by even a small all-round increase in consumption per head will place a considerable strain on world supplies. To the extent that these countries are able to satisfy the increased demand from indigenous energy resources, the effects on other energy-importing countries may not be too serious. On the other hand, if rising energy consumption in Third World countries has to be met out of the world's present exportable stock of energy, the impact could be very serious.

I have done no more here than point out a possible future danger. What are the chances of countries with low usage per head moving up to the level of the major industrialised countries? Japan's example has shown what can happen, and if India, for example, with her vast population, were to move up to Japan's present level of energy consumption per head, the effect on world demand would indeed be considerable. We must remember, however, that Japan's dramatic growth was based on cheap oil – an assumption which is no longer valid.

This is a very important factor in the situation and although nobody can foresee the future with confidence, the chances that another Japan-type transformation will occur in any country which lacks adequate indigenous energy now seems remote.

An enormous increase in China's use of energy certainly cannot be ruled out, but if it occurs it will be based on a corresponding development of indigenous energy supplies and the world balance should be largely unaffected.

Much more potentially serious is the likelihood of a further huge increase in *per capita* consumption in, say, Iran where the figure has already increased from about 0.1 tce in 1950 to 1.3 tce in 1974. A continued rapid increase in Iran's own consumption and that of other states similarly placed must be expected, and allowed for in our forecasts. Any such increases in consumption per head will cut into the energy surpluses which the countries concerned are able to export and on which the world is depending, as long as they last.

Call for Conservation

As I showed in the previous chapter, conservation policies were introduced in many countries during the years following the 1973 crisis and further steps of this kind will be increasingly applied in the future when they will exert a considerable influence on demand for energy.

Conservation measures can be of three different basic types, though the categories are not exclusive, since each interacts on the others. Into the first category fall energy savings due to the introduction of more efficient energy-using equipment. The second category includes savings resulting from action by energy users in response to higher energy prices and the third class of energy savings arise from specific government action.

It is the combined effect of all three types of measures which will affect future demand for energy as they have also contributed to the relatively low growth of energy in many areas over the four years since 1973. How much has been due to energy-saving equipment and techniques, how much to higher prices and how much to government action cannot be precisely determined, especially as a further factor pulling down energy consumption has been the economic recession.

Although direct government action is only one of the three factors on which energy conservation depends, it is the one which is most obvious to energy consumers and it also affects the other two. For example, where governmental action takes the form of taxation which makes any form of energy, or all energy, more expensive, the consumer will take action to use less energy, perhaps by installing new equipment. On the other hand, government action may provide a financial encouragement to take energy-saving measures.

Governments will clearly continue to take action on energy conservation, probably along lines similar to those already taken. As the seriousness of the world's energy problems become more obvious, the legislative action requiring people to take specific energy saving measures will become more stringent and the encouragments to save energy, by grants and other inducements, will become more attractive.

This combined carrot-and-stick approach will have a beneficial effect within the countries where it is applied. Such policies will also be helpful in an international sense, to the extent that anything which decreases internal consumption of imported energy in one country helps to conserve the world stock of that energy source.

Since all conservation policies involve some limits on the freedom of choice of individuals they must be restricted to what the "traffic will bear" and this will apply with even greater force to the international measures which will be required to deal with what is essentially a world-wide problem. Nevertheless, I am confident that international action on conservation will be stepped up, following on the initiative of the International Energy Agency (IEA) which produced a report on the subject in 1976, the influence of the OECD whose *World Energy Outlook* included a chapter on energy conservation and the work of the international Workshop on Alternative Energy Strategies (WAES). Successful conservation measures could have a considerable effect on future energy demand.

Measures which can be taken, under one or another of the three categories I listed at the beginning of this section of the chapter, are legion, covering almost every aspect of industry, agriculture, commerce and everyday life, from electricity generation, industrial plant and transport to entertainment and home usage.

Although it is easy to see how particular conservation measures will affect future demand it is never a simple matter to measure them in therms, tons of coal equivalent or any other convenient yardstick. However, estimates can be made of possible savings which will be sufficiently realistic to enable us to update demand forecasts as and when significant new energy economy measures are introduced.

Conservation alone will not solve the world's problems but it could, and I believe will, prove to be an important complementary factor to the other policies which the world must put in train as soon as possible.

Price Problems

Let us now consider the likely effects of price rises on future world requirements for energy. Price changes can have a double or even triple effect on the energy balance. People react to higher prices by trying to use less energy and so reduce demand on available supplies. Normally the cut in demand will be less than the increase in price which means more has to be paid for less energy, so that people have less to spend on other things, which in any case will cost more because of the higher cost of the energy content which goes into them. The net result will be lower economic growth and a further brake on the growth in energy demand.

A second effect of rising energy prices, however, works in the opposite direction and can increase the supplies of the various fuels coming onto the market. This is because, at the higher prices, it becomes economic to exploit resources which do not pay their way when prices are lower. For example, an oilfield, on or offshore, which is not regarded as being "commercial" when prices are at a given level, could be worth developing when prices rise to a higher level. Similar arguments can be applied to all the other fuels. Coal mines which were not previously economic become profitable with higher prices, while gas which is produced in association with oil and is normally flared away to waste can be economically recovered and transported to distant markets when prices make it worth while.

These two effects of rising energy prices react on one another and the end result will depend on the extent of the price rise, the availability of the reserves which can be brought within the "profitability bracket" and the speed with which they can be produced in response to the higher prices. This is a particularly important factor, because of the considerable and inevitable time lags which are a feature of most energy development.

Price rises are brought about in various ways and not only by shortages of energy or a fear of future shortages. Up to the present, the price rises the world has experienced have been mainly due to government action, more specifically the decisions of the OPEC countries. Eventually, however, oil prices will be rising, by the time the "real" crisis arrives, not only as a result of action by governments in exporting countries (who, by then, may well wish to restrict production to conserve their greatest assets) but also due to the additional supply-and-demand effects of diminishing reserves world-wide.

Energy price changes will thus exert a very considerable effect on future demand for energy, along with changes in economic growth and the results of conservation measures. Governments can exert some influence of energy prices, within their limited sphere of action, by legislation and policies, both commercial and fiscal and they will undoubtedly continue to do so. In the long run, however, it will be the hard facts of the world energy supply and demand situation which will set price levels and thereby influence future demand.

An Estimate of Demand

Having looked at the factors which will influence future energy demand, the reader is entitled to ask for an indication of the extent to which demand is likely to expand between now and the end of the century.

Forecasts of future energy demand in the past have often been wrong. Nevertheless, policies for production and conservation must be based on a view of the likely trend of future demand.

Almost all recognised authorities agree that, in order to achieve a balance between supply and demand, world oil prices are likely to rise a lot higher during the remainder of the century than has already been the case since the 1973 crisis.

In dealing with prices I shall be referring to real prices. Adam Smith, in *The Wealth of Nations* (1776), gave an explanation of "real terms" as they related to wages which remains as good a definition as any. "The labourer," he wrote, "is rich or poor, is well or ill rewarded, in proportion to the real, not the nominal price of his labour." In other words, the labourer's real wages were measured in terms of the actual goods his money would buy and not their value at current prices.

When applied to prices, "real terms" are thus a measure of the true value of the commodity under consideration, in this case oil, as related to the quantity of other products which a given amount of oil will buy. Inflation raises the prices of all products. A "real" price rise comes on top of price increases due to the general level of inflation.

Just how far energy prices will rise, apart from inflation, by the end of the century is naturally a matter of argument and speculation. A cross-section of informed opinion suggests that world oil prices will rise progressively to double the present level by 2000 – in real terms, of course.

Taking this as a basic assumption, in this and subsequent chapters I shall be examining the prospects of energy supplies being sufficient to match the demand if prices do double in real terms over the next twenty years or so. Although nobody can claim to tell the future in our changing world, and there are a great many uncertainties, the examination points strongly to the conclusion that, on the evidence available at present, the supply of energy will be likely to fall progressively below the level of demand at the prices I have suggested, unless far greater efforts than are at present in prospect are made to stimulate energy production and hold down demand.

Without this action, oil prices are likely to rise even faster than assumed above. They could rise to two-and-a-half or even three times their present level, again in real terms, by 2000 if corrective action is inadequate. On the other hand, really determined action world-wide to increase energy supplies and conserve energy by more efficient usage would probably result in the achievement of a balance between supply and demand with a much smaller price increase than our assumed doubling by 2000.

A balance will thus be struck somewhere between these two extremes and, with so many variables, it would be unwise even to hazard a guess as to the precise point where the balance will settle down, in 2000 or any individual year. I merely wish to underline the importance for the industrial world of not allowing any slackening of effort.

The more energy prices are held down by developing additional sources of supply and by applying effective forms of energy conservation, the greater will be the potential for higher living standards in real terms.

Let us now develop this thought to enquire what it may mean in terms of energy demand for the various fuels, in order of magnitude. (See Table 6).

Returning to our assumption that the world price of oil will rise in real terms progressively to double its present level by 2000, and assuming also a reasonable degree of world economic growth with a continuation of moderate success with energy conservation practices, energy demand in the non-Communist world seems likely to be around 9000 mtce in 1985, rising to, say, 14,000 mtce by 2000. These estimates should be compared with the actual consumption of 6150 mtce in 1974.

Though these prospective rates of growth are much lower than during the pre-1973 period, the figures nevertheless represent an enormous additional demand for energy by 2000, more than double the 1974 total, to be achieved in just over 20 years. Nor does this give the whole picture. The so-called non-energy uses of oil, notably in the petrochemical industry, itself looking for considerable expansion world-wide, are not included in United Nations statistics on which my figures are based. Oil for non-energy uses will have to come from the same sources as other supplies. Non-energy uses can be expected to add about 5 per cent to the projections of world energy demand in the non-Communist sector.

Increases in oil prices on the scale suggested will stimulate demand for other fuels to replace oil in applications where this is possible. How far the change in the pattern of use of the various fuels will go is likely to depend not only on their relative prices but also on availability.

Tentively, it will be assumed here that total production of the non-Communist world in 1985 of fuels other than oil will comprise, in round terms, 1800 mtce of solid fuel (against an actual 1224 mtce in 1974), 1750 mtce natural gas (1242 mtce in 1974), 650 mtce nuclear (83 mtce in 1974) and 600 mtce hydro-electricity (454 mtce in 1974). Oil would then be called on to meet 4200 mtce (against 3145 mtce in 1974) out of the overall energy demand of 9000 mtce.

To reach the overall total of 14,000 mtce by 2000, oil might have to attain a total of 5600 mtce, with solid fuel providing 2500 mtce, natural gas 2400 mtce, nuclear power 2600 mtce and hydro-electricity 900 mtce. These oil requirements are formidable indeed, and, as will be seen in later chapters, there must be great doubts as to whether supplies in the year 2000 will be high enough to meet these demands. There is a somewhat better chance of meeting 1985 requirements, although even that cannot be taken for granted. As things are going at present, the prospects of any deficit being made up by extra supplies of other fuels are far from good.

Table 6. Tentative Forecasts of Energy Demand in Non-Communist World (mtce)

	1974	1985	2000
Coal and lignite	1224	1800	2500
Oil	3145	4200	5600
Natural Gas	1242	1750	2400
Nuclear power	83*	650	2600
Hydro-electricity	454*	600	900
Non-Communist World Total	6148	9000	14000

*The conversion factor used for these fuels reflects their relative importance and is not that used by the United Nations Statistical Office.

To discuss the way in which a total demand for energy is made up perhaps gives the impression that there is complete interchangeability between different fuels but, although there is considerable flexibility in use, there is by no means full substitution between the various forms of energy.

With the present state of technology oil for example, is virtually unchallenged in the transport field and undeniably so in air transport. This could, of course, change if oil can be produced from coal (which is technically possible and on which I comment in chapter 8) on a widely competitive basis. On the other hand, all the fossil fuels can be used for electric power generation, for industrial process heating and for space heating if required. Consumer choice will generally be made having regard to convenience, cost and efficiency in use.

Changes from one fuel to another can seldom be made overnight, however, and a changeover on a scale likely to affect the world energy situation may take years, especially where it entails the large-scale installation of new plant to consume a new or an alternative fuel. The question of cost-effectiveness is vital here because the cost, in energy terms, of manufacturing and installing the new plant may exceed the energy savings achieved.

In the long term, however, and it is with the long term that we should be primarily concerned, there is likely to be sufficient interchangeability between the fuels to enable all resources to be used to the full, if required. We can therefore continue to consider whether the fuels expected to be available will meet total demand for energy, without being unrealistic.

I have given figures for demand in non-Communist countries rather than the whole world because the energy economies of the Communist countries are largely independent of those of the rest of the world. There are limited net exports of energy from the Communist countries to the others. The amounts of exports involved are only about 2 per cent of total non-Communist energy requirements, and are expected to remain at that level.

Many of the problems which face us in the energy field arise from the disparities between supply and demand from one area to another. It is not enough, therefore, to look at world demand in total but it must be broken down between the main areas, so that an impression can be gained of how world trade in energy will have to develop. The area-by-area composition of the 9000 mtce total energy demand of the non-Communist world in 1985, and the 14000 mtce in 2000 in round figures, could perhaps be divided as follows:

Table 7. Tentative Forecasts of Energy Demand By Area (mtce)

	1974	1985	2000
Western Europe	1726	2400	3600
United States	2564	3500	5200
Japan	479	800	1400
Other non-exporting areas ⱡ	1093	1700	2800
Exporting areas ∅	286	600	1000
Non-Communist World Total	6148	9000	14000

ⱡ Includes Australia and Canada, which are net exporters to a minor extent.
∅ Mainly OPEC.

Obviously too much significance should not be read into these particular figures. Nevertheless, they should be adequate for our purpose here, namely, to enable us to illustrate the energy problems facing the world in a practical fashion.

They are broad estimates prepared on the assumptions already specified (a doubling of oil prices by 2000, reasonable levels of economic growth and a moderate degree of energy conservation). Although available forecasts by recognised authorities have been taken into account in their preparation, these estimates do not represent the official forecasts of any country or group of countries, partly because official forecasts tend to be targets rather than best estimates.

As I have already said, much will depend on the extent to which the world heeds the plain warnings which many of us in the energy industries have been giving in recent years. Determined international action would reduce demand below the levels shown above, though it is not easy to make a reliable estimate of likely achievements. The OECD study *World Energy Outlook* showed that by adopting its "accelerated policy case", there could be a cut of 4 per cent in energy demand by 1985. This suggests that a proportionately greater reduction – say about 10 per cent – could be achieved by 2000.

If such a reduction in demand, allied to an accelerated programme of energy supply should take place, it would be more than sufficient to close the gap between supply and demand, on the assumption of a doubling of real energy prices by 2000.

Britain's Future

Future energy demand in Britain is an important element in the total picture but we should look at the British position separately because our energy prospects for the next 15 to 20 years differ materially from those of Western Europe generally, because of North Sea oil and gas, and our substantial exploitable coal reserves.

Official and unofficial estimates of the possible trend of demand for energy in Britain during the rest of the century show a wide "spread" between the upper and lower limits. It is not my intention to try to estimate where demand will fall between these limits. For my purpose here, it will suffice to have a rounded estimate for the UK on the same assumptions as those employed throughout this chapter.

On these assumptions, having regard to real price changes, economic growth and energy conservation, in very broad terms, it could be that demand for energy in the UK could be 400 mtce in 1985 and 500 mtce in 2000 (compared with 324 mtce in 1974 and 317 mtce in 1976).*

*Readers familiar with other estimates, such as those in the Department of Energy's *Coal for the Future*, should note that the figures given here are on the adjusted United Nations basis, as generally used in this book for the world picture and they cannot be directly compared with the Department's statistics.

There is, as we shall see, a fairly wide choice of how demand might be divided between the fuels in the intermediate years, but the degree of choice is likely to be narrowing off by 2000. As everywhere else, the demand for the various fuels will depend partly on price movements and partly on availability while the time scales necessary in practice to make changes in equipment to any great extent will also be an important factor.

Once again, this points to the need for early decisions on vital matters affecting future energy supplies.

I have tried to show in this chapter the scale of future energy demand which our policies must be designed to satisfy. In the next two chapters I discuss, first, the world's energy reserves, and, secondly, the way in which they are distributed around the world. This will provide an indication of the size of the world's present "energy bank" on which we must draw to meet future "energy bills", and help us to determine when energy bankruptcy may be upon us if we do not take the necessary avoiding action.

5 The World's Energy Reserves

As I implied in the previous chapter, the making of forecasts of energy demand is fraught with all the difficulties associated with any attempt to foretell the future in a rapidly changing world and there will be many, no doubt, who will disagree with the picture I have drawn. The disagreements, however, will I am sure, be only a matter of degree and will not invalidate my main thesis that the trend of demand will be upwards, (though at a slower rate than in the previous two to three decades) that sooner or later oil and gas will no longer be able to sustain the expected rate of growth, and that even with a substantial expansion in the output of the particular fuel which is available in abundant quantity, namely coal, the world energy situation will probably get more and more difficult.

Exactly when that situation is reached – toward the end of this century is the best estimate that can now be made – will depend, not only on the growth rate of demand but on reserves available "in the ground" and the way in which we exploit these reserves, and on such other forms of energy as may be open to us, from nuclear power to solar, wind, wave and geothermal energy.

In this chapter I shall be taking stock of the world's resources of energy as a whole – a problem which is no less difficult than the exercise in forecasting of demand undertaken in the previous section.

There are two reasons for this. In the first place, information obtainable from one source can rarely be compared directly with that from another. Definitions and units of measurement differ widely and there is no wholly satisfactory way of comparing various fuels as sources of energy. Secondly, there is the problem of assessing the quantities of coal, oil, natural gas and other resources which are at present under the surface of the earth or probably more important still in the future, under the sea. I shall deal briefly with these two problems before going on to summarise the latest information on the world's energy resources, presenting both the global totals of each fuel (in this chapter) and the main areas where the major reserves are located (in the following chapter).

It will then be seen that the largest reserves are not necessarily in the places where they are most needed and at the end of the appraisal I shall be discussing the prospects for increased international trade.

A Matter of Measurement

Without going too deeply into the somewhat complicated business of definition and measurement of all forms of energy, I feel I should perhaps offer a plain man's guide to the subject at this juncture, without which it is difficult to take an informed interest.

Coal, for example, is measured in tons but even this unit has at least four different values which can make a considerable difference to totals running into millions of tons. Traditionally, Britain has used the long ton (2,240 lb. or approximately 1016 kilogrammes) while the United States has the short ton (2,000 lb. or 907 kilogrammes). The metric ton or tonne (1,000 kilogrammes or approximately 2,205 lb.) is now gaining worldwide acceptance, including in the UK. The EEC also uses tonnes of "standard" coal (see page 56).

Oil is also measured in tons (long or short), in tonnes or, by volume, in barrels. Converting barrels of oil to tons or tonnes and *vice versa* is frequently necessary because newspaper reports and other sources of information for the general reader seem to use one or the other unit of measurement indiscriminately. Conversion is difficult, however, because the specific gravity of oil varies considerably – usually there are between 6.6 and 7.9 barrels to a tonne. However, rather than calculate the precise factor for each type of oil in the total under consideration, it is usual to employ an average conversion factor, such as the 7.33 used in BP's published statistics.

Gas is measured by volume, in cubic metres or cubic feet (1 cubic metre = approximately 35 cubic feet), but gas consumption is often recorded in heat units (e.g. therms or calories) based on the known characteristics of the gas supplied. Electricity capacity is measured universally in Watts (W), or a multiple, and electricity consumption in kilowatt hours (kWh) or a multiple. 1 kWh equals 1kW (kilowatt or 1,000 Watts) used for one hour. The kilowatt hour is a unit of energy and can be compared directly with other heat units.

When considering energy reserves from all these sources, it is often necessary to find a convenient means of expressing them in a common unit. Moreover, since the fuels are interchangeable to a considerable extent, a common unit is useful as a basis of comparison. This is usually achieved by comparing the amount of heat energy in the respective fuels, which can be measured in British Thermal Units, in therms or in calories. A British Thermal Unit (Btu) is equal to 252 calories and a therm is equivalent to 100,000 Btu's, or 25,200 kilocalories. Alternatively, different forms of energy can be measured by comparison with the heat energy of one specified fuel, usually, as we saw in chapter 2, in "tons of coal equivalent" (tce).

Unfortunately, different types or grades of the same fuel may have very different calorific (or heat energy) values. In Britain, for example, the heat energy in coal can vary from about 220 to 350 therms per ton while, on the world scale, the variations are even wider, especially when lignite (a low calorific value fuel also known as "brown coal") is included. The EEC equates 7,000 calories to one gramme of standard coal (hence 7 million kilocalories = 1 tonne of standard coal) which means 1 standard tonne of coal is about 278 therms.

Oil varies widely from one product to another, ranging roughly from 400 to 500 therms per tonne, as well as varying appreciably from one crude oil to another. North Sea natural gas has a calorific value about twice that of the manufactured (or town's) gas it has now replaced in the UK, as anyone knows whose gas appliances have been converted to the newer fuel. There is, however, a constant relationship between the unit of electricity and a therm (29.3 kWh = 1 therm).

A precise heat content for coal and oil is rarely recorded and an average conversion factor is commonly used. Different authorities and organisations use different conversion factors and even the same authority may use different conversion factors for specific purposes.

As an example of differences in conversion factors, the European Economic Commission treat 1 tonne of crude oil or oil products as equivalent to 1.43 tonnes of standard coal, the United Nations in their latest statistical tables (Series J No. 19) regard 1 tonne of crude oil as equivalent to 1.47 tonnes of coal and 1 tonne of an oil product as 1.50–1.61 tonnes of coal, depending on the product, while in Britain the Department of Energy equates 1 tonne of petroleum products to 1.7 tonnes of coal.

A special difficulty arises when converting primary electricity (that is, hydro or nuclear power) to coal (or oil) equivalent. It is possible to compare primary electricity with coal of the same heat content and this is done, for example, by the United Nations, which equates 1000 kWh to 0.123 tonnes of coal. Since it takes much more coal than that to produce 1000 kWh of electricity (and correspondingly more oil or natural gas), most other authorities equate primary electricity to the approximate amount of coal that would be required to produce it. The EEC, for instance, in 1975, assumed that 1000 kWh was equal to about 0.32 tonnes of coal (although their figure varies from country to country and from year to year).

From all this it will be only too obvious that the measurement of energy is a complex business and, although I am not suggesting that the general reader need be too concerned with the minutiae of the statistical problems, it is well to know that these problems exist. Otherwise, it is all too easy to be misled when making comparisons between figures drawn from different sources which may not use the same definitions and conversion factors.

In the hope that this note on terminology will prove useful, not only when reading later sections of this book, but also when seeking to understand the implications of published information, I now pass to an equally complex matter – the methods used for measuring actual reserves of fuels.

Assessing Resources

One of the questions which is most frequently asked is: "How do we know how much coal, oil or gas lies beneath the earth's surface or the seabed?" The answer, of course, is that we cannot be absolutely certain of the precise content of any particular deposit, but neither are the figures quoted to be regarded as merely guesswork. On the contrary, up-to-date techniques of exploration have reached the point where we can obtain estimates of reserves which give a sufficiently accurate picture on which to base commercial and technical decisions.

Techniques used in exploring for coal, oil and natural gas have been brought to a pitch of technical sophistication which would require a complete chapter to describe in detail and is outside the scope of this book. However, the seismic, magnetic and gravimetric techniques now in use enable mineral surveys to be conducted with greater accuracy than before, though for really reliable estimates of reserves, full-scale boring programmes are essential. These are expensive, both on shore and still more so when exploring offshore reserves – in the North Sea the cost of a single exploratory well exceeds £1.5 million and it is common for ten such wells to be drilled to delimit a major discovery.

It follows, therefore, that boring is only undertaken when there seem to be good prospects of profitable production on the basis of preliminary investigations. This is certainly our policy in the British coal industry and, as we shall see, it paid dividends in the discovery of the Selby coalfield in Yorkshire.

For the same reasons, reliable estimates of reserves of coal, oil and natural gas are only available world-wide for countries and regions which are either being or are shortly to be commercially exploited. Estimates for areas which have not been tested to this extent are usually made by analogy with similar, fully tested areas elsewhere and the margins of error can be large. The distribution of fossil fuels varies widely – even in areas which otherwise have similar characteristics. There remain considerable areas of the world which are as yet largely unmapped geologically and totally unexplored by prospectors and here only the broadest guesses can be made. Energy reserves below the oceans are even less well known.

Since even the reliable estimates are made with commercial exploitation in mind, those responsible are more concerned with the amounts which can be recovered economically under the prevailing technical and commercial conditions than with the absolute quantities in the ground. As economic conditions change and technology advances, the estimates of recoverable reserves may change accordingly, without any change in the basic information as to what lies below the surface.

Because of the different ways in which reserves of all fossil fuels are estimated, they are conventionally broken down into three levels of certainty, respectively representing reserves which have been measured (by the methods described), indicated by relevant information or merely inferred. These three levels are more usually described as "proven", "probable" and "possible" reserves (in descending order of likelihood), and it is important to note which term is being used when discussing reserves.

Sources of information on reserves

Estimates of the world's total energy reserves are published from time to time but need to be regarded with considerable caution, not only because of the big variations in the quality of information available from different regions, but also because there are no internationally-agreed conventions for recording information on reserves, apart from the general terms mentioned above.

No international body, as yet, attempts to produce annual world-wide estimates of reserves of all the main fossil fuels (coal, oil and natural gas). However, a periodical survey of world energy reserves is produced by the World Energy Conference (previously called the World Power Conference).

This is an international non-governmental body, with national committees in the various member countries. These committees collect such information as is available in their respective territories and this information, plus other published data, forms the basis of the survey of world energy resources produced by the Conference. The survey deals both with known reserves and with the total resources believed to exist, country by country and fuel by fuel.

Apart from the obvious variations in definitions and measurements employed in the different countries, the reliability of the information also varies and there are a number of countries which either supply no data at all or provide only limited information.

Surveys conducted in this way by the World Energy Conference have appeared at 6-yearly intervals since 1962. The 1974 Survey, which appeared in conjunction with that year's Conference in Detroit, attempted a considerably more detailed approach than previously, both quantitatively and in the quality of the information presented.

It was followed up by a 1976 Survey giving certain revised and updated data, and further estimates were given at the 1977 Conference in Istanbul.

For oil and natural gas annually updated information about the world's proved reserves is, however, issued by the *Oil and Gas Journal* which gathers information from sources all over the world during the year and is regarded as an unusually authoritative source. The American Petroleum Institute and the Canadian Petroleum Association also issue regard annual assessments of reserves for the countries they serve.

These latter sources, together with the *Oil and Gas Journal*, form the basis of the figures for oil reserves given in the *BP Statistical Review of the World Oil Industry*, which is published annually and is recognised as the best available indication of known oil reserves, certainly in the UK. However, for annual figures of natural gas reserves, country by country, world-wide, we have to rely on the *Oil and Gas Journal* figures alone.

No annual figures of coal reserves world-wide are published, although periodical reassessments of national reserves are certainly made by the countries concerned, not necessarily for publication. We in the UK keep our reserves under constant review, as I shall explain fully in the next chapter.

Anyone wishing to obtain an up-to-date picture of the world's reserves of all fossil fuels, therefore, must make do with such positive information as is available. Bearing all these reservations in mind, I summarise here the latest information available.

The 1975 edition of the *BP Statistical Review* included a diagram which compared proved recoverable reserves of the fossil fuels, in terms of energy content. Measured in thousand million tons of oil equivalent the approximate figures were crude oil, 90, natural gas, 44, tar sands, 12, oil shale, 29 and coal 343.

In the UK, the Department of Energy, in a paper entitled *World Energy Resources: Present Position and Prospects*, brought together recently published figures of total recoverable reserves, including probable and as yet undiscovered resources. Not unexpectedly, the totals are far higher than the BP Statistical Review figures, which referred only to proved recoverable reserves. The Department of Energy's figures, also in thousand million tons of oil equivalent, showed oil, 233, gas, 171, coal and lignite, at 10 per cent recovery rate, 645 or, with 50 per cent recovery rate, 3225.

Both sets of figures emphasise the overwhelming importance of coal, even though there is considerable uncertainty about how much of these reserves will be economically recoverable, especially in the remoter areas of the USSR. This is why two different recovery rates are quoted. However, even on the more conservative assumption, coal is three times as plentiful as oil.

The World's Oil

For oil, most published estimates relate to proven reserves, such as the BP Statistical Review figures. The 1976 Review showed that world reserves stood at 88.3 thousand million tonnes at the end of that year, representing about 30 years' supply at 1976 rates of production. This estimate is not very far removed from the 91.6 thousand million tonnes of total proven recoverable reserves given by the 1974 World Energy Conference.

The oil companies and the various authorities concerned in presenting figures for petroleum reserves are normally, though not always, fairly cautious in deciding what is "proven", so their figures are likely to be uprated in the future as deposits gradually move up the scale through the "possible", "probable" and "proven" categories. Higher prices and advancing technology will also increase the amounts that are recoverable, and there remain some regions of the world which are as yet totally unexplored for mineral wealth.

The international oil industry is turning increasingly to the less accessible oil reserves known to lie offshore on the continental shelves in many parts of the world, notably Latin America, South East Asia, the Middle East and India. Growing attention is also being paid to the less accessible reserves in such inland areas as the Arctic islands of Canada, Alaska and the Trans-Andean basins of South America. Oil from some or all of these areas will gradually figure in the "proven" category, but any oil from these areas will be far from cheap.

During the 1960's proved oil reserves continued to advance ahead of production and the world seemed to assume that this could go on for ever. By the early 1970's authoritative voices were warning that this was not so, and that oil resources were relatively limited. Harry Warman, for example, then chief geologist of BP, gave an estimate of total recoverable world resources of oil which has been widely accepted – about 270 thousand million tonnes. He showed that on this basis a growth in world oil demand at 7½ per cent a year – as in the 1960's – could not be sustained beyond the end of the 1970's, and that world oil production would reach a peak of around 6 thousand million tons a year in the early 1990's, from which it would gradually decline.

Now, of course, with oil prices so much higher, nobody expects a growth rate of 7½ per cent a year. But even with no growth in demand at all, the total resources of 270 thousand million tons would not support oil production at present levels for more than a few years (less than 10) beyond the year 2000, and a slow decline would then set in. We saw in Chapter 4 that demand for oil was in fact expected to continue to grow, unless governments throughout the industrialised world took decisive steps to stop it, or prices were raised so high as to choke off growth.

I have stressed the uncertainty of any estimate of total world resources of any particular fuel. Yet even if total recoverable resources of oil prove to be half as high again as Harry Warman's estimate, the year of peak production could only be 10–20 years further off. A consensus view of top authorities given in a paper presented to the 1977 World Energy Conference in Istanbul – six years after Warman made his estimate – came up with almost precisely the same figures as he did.

There is another, and possibly better, reason behind the assumption that the world's oil resources will not support the possible demands for oil much longer. Discovery rates of new oil reserves have been disappointing in recent years, and it is widely believed in the oil industry that they are likely to decline because the easiest and most accessible reserves have already been discovered. Robert Belgrave of BP has suggested that discovery rates by the end of the century will have been almost halved. He concluded that this might cause oil production in the non-communist world to peak at a level only 50 per cent higher than in 1974 – appreciably below the level of demand in the latter years of the century which most authorities expect.

One of the few leading figures in the energy debate who does not expect world oil production to have started to fall before the year 2000 is Professor Peter Odell of the Economic Geography Unit at Erasmus University, Rotterdam, [appointed a part-time consultant to the U.K. Department of Energy in October 1977] who is well-known for his criticisms of conservative forecasts of oil reserves. Whatever the merits of his case, the weight of the evidence suggests that even significant upgrading of oil reserves would do little to weaken the central thesis in this book, namely, that within a relatively few years oil will become progressively more scarce and very much more costly and that we should press ahead now with plans to make up the deficit by other means.

Natural Gas

Continuing the survey of world energy resources, we come to natural gas which, until comparatively recently, was not generally regarded as one of the major energy sources, except in a very few countries of which the United States was the best example. Natural gas is often, though by no means always, found in association with crude oil and, although it has sometimes been harnessed as a by-product of the oil industry, it has been flared wastefully away in many cases.

One reason for this apparent thoughtless waste of potentially valuable fuel is that natural gas is much more expensive to transport than oil, unless found near a point of demand. Delay in making use of the world's natural gas resources has been partly due to the transport problem but the development of new long distance pipe-laying techniques, especially in the United States after the Second World War, and, more recently, the transport of natural gas in liquefied form (LNG) by sea, has changed the situation. The five-fold increase in oil prices has provided additional encouragement, though much of the under-utilised reserves are in countries which are already large-scale exporters of oil. These countries are hardly likely to allow sales of natural gas to weaken the price of their oil.

Over the past 25 years, the world gas industry has grown to nearly seven times its former level, compared with a growth of five times for oil. Today, natural gas provides nearly 20 per cent of world energy production, although it constitutes no more than 8 or 9 per cent of the world's total proved reserves of fossil fuels.

An accurate assessment of world natural gas reserves is no less difficult to obtain than for oil. Published proved reserves of natural gas are issued annually by the *Oil and Gas Journal* and these figures are generally accepted as the best available source of information on known world reserves. The world total at January 1st 1977 was put at 2,327 trillion cu. ft., equivalent to about 55 thousand million tons of oil, which suggests that natural gas reserves are around 60 per cent of world reserves of oil.

Exploration for natural gas in the world as a whole has, up to now, been much less intensive than for oil. In the United States and Canada, where both oil and gas production have been extensively developed, published proved gas reserves are higher than the corresponding figures for oil. This may suggest, though by no means conclusively, that when the rest of the world has been equally thoroughly explored, the eventual resources of natural gas may prove to be considerably more than the 60 per cent of oil reserves mentioned above.

This view is implied in the UK Department of Energy paper *World Energy Resources: Present Position and Prospects*, where eventual reserves of natural gas are put at about 75 per cent of those of oil. Yet whatever the total resources turn out to be, the supplies will eventually run out, following the pattern already foreseen for oil. Moreover, much of the gas will be extremely costly to produce and transport to the centres of demand. The Department of Energy's paper shows that if world consumption grows exponentially at four per cent a year, natural gas resources will last theoretically for about 50 years, though the stage of declining production will be reached long before then.

The Biggest Energy Source

Coal has undeniably the biggest reserves of all the world's major sources of energy and is therefore the brightest hope for the future, if only its potential is realised and the necessary action taken now. The question is how much of the huge quantities believed to exist in the ground will prove economic to extract.

According to estimates presented at the World Energy Conference in 1977 the world's geological resources of solid fuels amount to more than 10,000 thousand million tonnes coal equivalent. It was reckoned that some 640 thousand million tonnes c.e. were technically and economically recoverable, under the conditions prevailing today. This is equivalent to between 200 and 300 years at current rates of usage (2.7 thousand million tonnes).

Nor is it even certain that the huge figure of 10,000 thousand million tons coal equivalent is the ultimate in world solid fuel reserves. Other authorities put it still higher. For example, as long ago as January 1st 1967, Paul Averitt of the United States Geological Survey, estimated the world's solid fuel resources of all kinds, as determined by mapping and exploration plus an estimate of reserves in unmapped and unexplored areas, at no less than 15,200 thousand million tons.

Much of the coal existing in the ground lies in regions far from the main centres of consumption and where the climate is inhospitable and the terrain difficult. A great deal of it is in seams that would be difficult if not impossible to work with present techniques. Nevertheless, it is evident that there is sufficient coal, world-wide, to meet any foreseeable demand indefinitely and to replace, for a very long time ahead, any other fuels which may be in short supply and for which coal is an adequate substitute.

The cost of extracting a great deal of these reserves will undoubtedly be high and the time scale long. In some of the regions there may be environmental obstacles to development. However, present-day notions of what is economically and technically possible and of the relative importance of environmental values and human needs may well undergo a radical change as the realities of the world's energy situation strike home.

This view appears to have been accepted by the international oil companies – who are shrewd observers of the energy scene – as their diversification plans now include the purchase of substantial stakes in the coal industry, notably in the United States and, more recently, in Australia and elsewhere. Since it was the world oil industry which caused the decline of coal during the late 1950's and 1960's, it seems that the wheel is now coming full circle, as the oil companies begin to invest in coal's future.

Primary Electricity

Coal, oil and natural gas – these are the three main pillars of the world's energy reserves, now and for the rest of the period we are considering. Other forms of primary energy include hydro and nuclear electricity. The former is subject to natural limitations, while nuclear power, at least as at present envisaged, depends on the world's reserves of uranium.

Unlike coal, oil and natural gas, however, hydro electric power renews itself year by year, though to a varying extent, depending on meteorological conditions. It follows, therefore, that once the world's hydro reserves have been fully harnessed, they will not face gradual depletion as will the fossil fuels but will continue at the same rate indefinitely.

Unfortunately, both the present and potential significance of hydroelectricity are small when compared with the major fuels available today and with the world's future energy needs. In 1974, for example, world consumption of hydro-power was reckoned to be 180 mtce, or only about 2 per cent of total energy consumption – admittedly on the United Nations basis of comparison which, as we have seen before, tends to undervalue the contribution of primary electricity.

Theoretically at least there remains considerable scope for further development, for the World Energy Conference of 1974 showed potential world hydro-electric resources of 9,800 thousand million kilowatt hours a year, or about seven times the present output of hydro-power. Since it would require over 3,000 million tons of coal a year to produce this amount of electricity in conventional power stations, the potential contribution is far from negligible. However, it is unlikely that anything approaching these maximum potential resources will be developed – at any rate within the time span with which we are concerned.

Hydro-electric power has to be produced where the hydraulic energy is available and the more remote the area the bigger are the problems of transporting the electricity to points of use. The problems of construction and operation of power stations in what are often wild and inhospitable regions are also considerable, and the costs of power generation under such circumstances very high.

At the present time, the best sites in the developed nations have already been exploited. Remaining sites will be increasingly costly to develop and operate. The as yet untapped sources in the already developed regions are therefore unlikely to make a major contribution to the future energy requirements of these countries.

Possibilities are far greater in the developing areas, where very large potential capacity is still unexploited. Hydro-electricity should be capable of providing a substantial fraction of the energy needs of Asia, Africa and Latin America, although the requirements of these areas are small compared with those of the developed world.

As with every other energy resource, the present stage of utilisation depends on what is economically and technically viable. As time goes on and the energy situation changes, sites which are at present regarded as uneconomic will become worthy of development.

Nuclear Power

Potential world resources of nuclear power cannot be assessed in the same way as resources of other fuels because so much depends on the extent to which, and the way in which the world decides to exploit this form of power; and it is at present subject to a good deal of controversy.

However the situation develops in the future, a dominant factor will be the world's reserves of the basic fuel uranium, which is consumed by almost all types of nuclear reactor currently in use or under construction. To that extent, therefore, nuclear power is yet another energy source which is finite.

A Report published by the Organisation for Economic Co-operation and Development (OECD): *Uranium: Resources, Production and Demand 1975* illustrates the complexities and uncertainties of the world's uranium resources. Present known reserves of uranium (exploitable at costs of not more than 15 dollars per lb U_3O_8) are given as no more than 1 million tonnes.

This would meet only about a quarter of the requirements expected in this Report to arise by the year 2000 – leaving nothing for the future life time of reactors in commission by that date. More than 80 per cent of these reserves (and of present production capacity) are in the United States, Canada, Australia and South Africa – a concentration of reserves which may restrict the availability of uranium for the world market.

Including rather more costly supplies (exploitable at costs of up to 30 dollars per lb U_3O_8), total resources of uranium known or estimated to exist were put at 3½ million tons by the OECD study. Large areas of the world remain to be explored and probably much more uranium could be discovered. Exploration in new areas however is not progressing fast and since it takes considerable time – perhaps as long as 15 years – from the start of exploration to initial production, these areas may make only a limited contribution during the rest of the century.

Uranium can also be extracted – at higher cost still – from low grade uranium ores. The OECD Report made no attempt to estimate how much uranium could be extracted from these low grade ores at costs which would enable nuclear power to compete with electricity generated from coal or oil. Whatever the theoretical availability, however, there is little sign so far that any substantial quantities of uranium will be produced from low grade ores this century.

A great deal will depend on the evolution of the fast breeder reactor. Without going into technicalities, the significance of the fast reactor (the experimental 250 MW reactor at Dounreay, on the north coast of Scotland, is of this type) is that it will allow a greatly increased amount of electricity to be generated from a given quantity of uranium. A recent Department of Energy Report showed the importance of the role of the fast reactor. Total recoverable reserves of uranium (including probable and undiscovered categories), assuming uranium prices up to 30 dollars per lb., are put at 59 thousand million tons of oil equivalent, if used in conventional thermal reactors but at 2,932 thousand million tons in fast reactors.

Clearly, therefore, the way in which nuclear technology is developed in the future will profoundly affect the duration of the world's uranium supplies. The OECD Report mentioned above estimated that by the year 2000 about 10 per cent of nuclear capacity would consist of fast breeder reactors.

Sir John Hill, Chairman of the UK Atomic Energy Authority, writing in the NCB Coal and Energy Quarterly (Autumn 1975), said: "The stocks of waste uranium stored in Britain would, if used in fast reactors, meet the present electricity demands of the country for several hundred years. There can be no real uranium shortage."

Although Sir John was referring to the domestic UK scene, much the same conclusions will presumably apply world-wide. However, the OECD estimate that by 2000 some 10 per cent of the world's reactors will be fast breeders looks decidedly optimistic, in view of the difficult technical, environmental and strategic issues still to be solved.

Alternative Energy Sources

Any survey of world energy reserves must make some reference to the resources which are as yet only at the early development stage and which will therefore be discussed in more detail when I consider the exploitation of the world's energy reserves. For completeness sake, however, I will mention these resources which include the shales and tar sands from which oil can be extracted as well as geothermal resources and the renewable forces of solar energy, wind and wave power.

Shales and tar sands exist in the Americas in large quantities, mainly in the United States, Canada and Colombia and there are smaller quantities elsewhere in the world, including a little in the UK where they are found in particular in Southern Scotland. The World Energy Conference of 1974 estimated known recoverable reserves of oil in shale and tar sands at 318 thousand million tonnes. The fact that the 1974 figure was more than three times the Conference's 1968 estimate suggests that new reserves are being discovered at a rapid rate. The new figure however, probably included considerable quantities which it would not be economic to recover.

BP's study of *World Energy Prospects* (August 1973) put world recoverable reserves (ultimately expected to be discovered) of shale oil at 435 thousand million tons and of tar sands and heavy oil at a further

33 thousand million tons. These estimates of ultimate recoverable oil from shales and tar sands can be compared with the estimate given in the same study of ultimate recoverable "conventional" oil at 220 thousand million tonnes.

An assessment of the world's total resources of oil in shales (that is, including possible and as yet undiscovered sources), undertaken by the 1974 World Energy Conference, worked out at the astonishing figure of 53,000 thousand million tonnes.

In spite of this potentially enormous source of energy, however speculative the precise figure may be, there does not yet appear to be a satisfactory commercial process for the large scale exploitation of shale oils and tar sands. Attempts to do so in recent years have underlined the great economic and environmental problems which will have to be solved. It is unlikely that, in the period covered by this book, any great part of the oil in shales and tar sands will have been tapped.

There is little point in attempting to give figures for the world's reserves of the other "unconventional" energy sources since everything depends on economic trends and technological advance. In a general sense, of course, the so-called benign sources could solve our problems. If the full power of the sun could be harnessed, for instance, it could provide for the world's total energy needs many times over. However desirable, I feel it is unrealistic to expect rapid advances in the next 20 years or so.

At one time in our history, wind and water power provided the greater part of our non-human energy, from driving mills to powering ships but although there are plenty of ingenious ideas for applying modern science and engineering skills to the utilisation of these forces of nature, I do not see them playing more than a marginal role in the world energy picture until well into the next century.

Geothermal energy (that is heat from the interior of the earth whether escaping naturally or extracted by drilling and pumping) is a large capital reserve of energy but only in relatively few places can it be tapped easily enough for economic exploitation. At the end of 1974, the total capacity of installed geothermal plant was about 1200 MW of which some 40 per cent was Italian and another 40 per cent in the United States, providing about 0.03 per cent of world energy needs.

It is interesting to speculate on the vast possibilities inherent in the forces of nature, could they only be harnessed. Writing in the *Scientific American,* back in 1971, M. King Hubbert reckoned that the solar energy driving the weather machine which causes wind, waves and sustains the hydrological cycle, is in the order of $40,000 \times 10^{12}$ watts. This can be compared with the world's total electricity capacity in 1974 which amounted to 1.5×10^{12} watts!

This is a long-term speculation. For many decades to come – and certainly beyond the end of this century – the world will primarily depend for its energy on fossil fuels. Their relative availability and accessibility will be crucial. In the next chapter I shall consider how the known reserves are distributed between different countries and regions, including our own.

6 Where the Reserves Are

A general appraisal of the world's energy resources, such as I have attempted to give in the previous chapter, is useful as an overall picture but it overlooks the fact that the world's major fuel deposits are unevenly distributed and not always in the places where they are most needed.

Very few countries can claim to be completely self-sufficient in energy. The Soviet Union is the most obvious example, though Britain hopes soon to be in that happy position, for a time, except for uranium. Some countries are self-supporting in one or more forms of energy, while others – Japan comes immediately to mind – are almost devoid of indigenous reserves.

Many of the world's recent energy problems arise out of this imbalance – the 1973 crisis is the prime example – and, for as far ahead as we can see, it will continue to pose problems. I therefore feel it is particularly important to know where the major deposits of the main fuels – coal, oil and natural gas – are located and I shall deal with each separately, leaving the United Kingdom out of the general survey, for separate, more detailed, treatment later.

To overcome the problems caused by the mal-distribution of energy reserves, it is necessary to transfer energy, in appropriate forms, from the energy "haves" to the "have-nots". I shall refer to the prospects for increased international trade especially in coal and natural gas, since the international movement of oil across the world is already highly developed.

Because the most urgent problems are associated with the world's oil resources, I begin with the distribution of this form of energy.

The World's Oil

A few simple statements will demonstrate the uneven distribution of world oil resources. Nearly two-thirds of total published proved oil reserves are located in the Middle East and Africa, with one-sixth in the Soviet Union and the other controlled economy countries, of which the lion's share is in the USSR. This leaves about one-sixth for the rest of the world, including the whole of the Americas, Western Europe (including the North Sea), Asia and Australasia.

Some more comparisons will further underline the extreme geographical imbalance of the world's oil and the worrying prospects facing the Western World. For example, whereas the proven reserves of the Middle East and Africa amount to some 42 years' production at present rates, the rest of the non-Communist world can only sustain about 18 years' production. The proved reserves of the whole of Western Europe, including the North Sea, at the beginning of 1977, were actually less than those of Abu Dhabi, one of the smaller members of OPEC.

On the most recent published figures of proved reserves, it would require 16 North Seas to match the reserves of the Middle East and North Africa. It is therefore self-evident that the industrialised nations, including Japan which has no oil resources of its own, must look to the Middle East and North Africa to supply the greater proportion of their future needs.

Oil is also unevenly distributed among the Middle East countries. Saudi Arabian official estimates of proved reserves are nearly 2½ times as high as those of Kuwait and Iran, who have the biggest reserves. It is a sobering thought that the prolific Middle East oil producing area, so essential to the world's energy economy, has now been extensively explored and any future major discoveries will probably have to be made elsewhere.

Western Europe is unlikely to make a very large contribution to the world oil supply position, since it has only 4 per cent of the world's proved oil reserves, even when the North Sea is taken into consideration. Britain's sector accounts for two-thirds of Europe's total; the only other European country with substantial oil resources being Norway, whose proved North Sea reserves are about one-half those of the U.K.

With two-thirds of the world's proved reserves of oil located in the Middle East and Africa it is clear that this area will continue to remain the main source of supply for countries without, or with insufficient, indigenous oil. As for the rest of the world's oil, a great deal will depend on whether those countries which possess reserves have sufficient for their own requirements will need to import some oil or can find a surplus for export.

With this in mind, I look at two big oil producers who are also large consumers – the United States and the USSR, both of which can profoundly affect future developments.

United States Oil

Until 1975, the United States could still claim to have the highest oil production rate of any country in the world, although, since then, both the USSR and Saudi Arabia have moved ahead. Even so, US oil production represents less than three-fifths of the country's total oil consumption. Proved reserves add up to about 11 years' production at the reduced 1976 rate.

A Report prepared by the National Petroleum Council in 1972, at the request of the United States Department of the Interior, showed how important it is to differentiate between "reserves" (defined as proved, recoverable oil) and "resources" (or the quantity in the ground, including oil not yet discovered, or considered not economically recoverable at present). On this basis, the 1972 Report estimated resources (also known as "oil-in-place") at rather more than 100 thousand million tonnes, or nearly ten times the published proved reserves of the United States and actually greater than the total published proved reserves of the whole world, at the end of 1976, as given by B.P. (88.3 thousand million tonnes)!

This emphasises the importance of price when considering oil resources. More than half the oil which remains to be discovered, as mentioned in the Report, is expected to lie in the frontier areas of Alaska and offshore, with very little left in some of the mature inland fields. Most of these reserves will be very costly to exploit, since much of the offshore oil, in particular, lies in conditions considerably worse even than the North Sea.

Much the same point was made in Circular 725 of the US Geological Survey, reported in the Federal Energy Administration's major study *National Energy Outlook, February 1976.* The Circular separated reserves and resources between those which were considered economic at 1973 prices and those which were not (described as "sub-economic"). The differences between the two assessments are striking.

As an example, economic reserves (at 1973 prices) were given at 4700 million tonnes, compared with uneconomic measured reserves of 16,300 million tonnes. Total reserves and resources (measured, indicated and inferred reserves and undiscovered recoverable resources), economic at 1973 prices, add up to 20,500 million tonnes. The same categories which are reckoned to be uneconomic at 1973 prices give a total of 26,800 million tonnes, or a grand total of oil-in-place of 47,300 million tonnes.

What was not reported was how much of the sub-economic reserves and resources would be economic at post-1973 prices, perhaps because this is something which only the future will show.

Even greater uncertainty surrounds the possible exploitation of the US oil shale deposits and oil sands. Shale deposits are located in several areas, including the Green River formation, in Colorado, Utah and Wyoming, which is much the most attractive in size and availability. This formation has been estimated to have an oil-in-place figure of at least 250 thousand million tonnes, although much of it consists of poorly defined and unfavourably located deposits, with low potential yields of oil per ton of shale. The more accessible and better defined resources were put at about 17 thousand million tonnes. These resources are not at present being exploited in quantity and to develop large-scale production would be difficult and costly.

At one time, US shale oil reserves were regarded as a potential major factor in the world's energy supply prospects, but the problems of developing them have proved far greater than anticipated. It is significant that the Federal Energy Administration's *National Energy Outlook February 1976* scarcely mentioned shale oil.

Tar sand deposits in the United States, though not on the same scale as shale oil, are by no means negligible. So far, however, they appear to have attracted little interest and, on the limited knowledge available, they are not expected to have any significant effect on the US energy situation, even by the end of the century.

Given appropriate financial incentives the US oil industry could probably achieve a higher rate of discovery of new reserves than in recent years, at least for a time, and step up production above recent levels. But there is no sign yet that we can expect the massive and continuing increase in the rate of discovery of new reserves which would be needed for oil production to be brought up to the level of consumption.

Though the United States may go some way towards meeting its future growing demand for oil, it seems likely that imports will remain on a relatively large scale. There will thus be a continuing demand on the same stock of exportable oil to which the rest of the world will also be looking. At present, imports of oil into the United States, at some nine million barrels a day, amount to more than three times the likely peak production rate from the U.K. sector of the North Sea.

Soviet Oil Resources

There is no doubt that the USSR has vast oil resources, especially in the Siberian region, where much of the country's natural gas and coal are also found. Although these great riches are known to exist, the transition from exploration to exploitation is only just beginning and precise information as to their extent is not readily available. Enough is now known, however, to enable an outline picture to be drawn.

Russia is no newcomer to the international oil business, having an old-established industry, primarily based on fields near Baku on the Caspian Sea, followed, after the Second World War, by further discoveries in the extensive Ural-Volga region. Many of these fields, however, have either reached or passed their production peaks and the search for new reserves has moved further east, into Western Siberia, where the first oil strike occurred in 1960.

Most of the main oilfields so far located – over 60 have been identified, of which about a dozen are being exploited – are in the Tyumen region of Western Siberia, roughly on a parallel with Leningrad, and well to the north of the main areas where economic development is proceeding. The largest oilfield (Samotlor) is credited with an eventual potential production of between 80 and 100 million tonnes a year. It is connected by a pipeline to refineries in the Ural-Volga area, over 1,240 miles away, with another pipeline to a big refining centre near Irkutsk, in Eastern Siberia.

It is believed that the Soviet Union still holds vast undiscovered oil reserves in its widely spread sedimentary areas. The Eastern Siberian region has not yet been thoroughly examined and there are other possibilities in Artic Siberia, although this area is bound to present extreme operational difficulties. In the Far East of the country there is some oil production in North Sakhalin but little is known of the full potential resources of the region.

If these resources could be economically exploited, it would no doubt open the way for exports to the countries of the West, but at the moment the USSR is having difficulties in meeting the calls of her East European neighbours.

Assuming that the USSR is unlikely to add significantly to the world's exportable oil stock and that the US seems certain to continue to be a net importer, Western Europe and the rest of the oil importing world will be forced to rely increasingly on the Middle East. What hopes for natural gas?

The World's Natural Gas

Since natural gas has only been extensively utilised worldwide in comparatively recent times, the distribution of reserves around the world is not as clearly established as in the case of oil and coal. The Groningen gas field in Holland ranks in sixth place among the world's published proved gas reserves, after the USSR, Iran, United States, Algeria and Saudi Arabia.

To put the UK North Sea reserves in perspective, they are about half the Dutch reserves while Norway's North Sea reserves are about 60 per cent of those in the UK. *Oil and Gas Journal* figures for January 1st 1977 show that the six countries listed above, plus Canada, all have bigger reserves than the UK and Norway combined. Nigeria, Venezuela, Australia and Kuwait, together with the "big seven", have bigger natural gas reserves than Britain alone.

As with oil, the USSR and US have an important role in the future of natural gas reserves, and both deserve separate consideration here.

Published proved reserves of the USSR are considerably bigger than those of the Middle East and North Africa combined and over four times those of the United States. Moreover, they are continually being revised upwards. They represent some 70 years production at present rates but, undoubtedly, when probable and possible reserves are included, the potential output of the deposits will be still further extended.

Because of the vast size of the USSR, the natural gas deposits are widely separated and – particularly in the case of Siberian reserves – remote from the main centres of population. This factor is all the more important because nearly all recent discoveries have been made in Western Siberia, where some 60 per cent of known reserves are located. A further 16 per cent is in Central Asia and Kazakhstan and 2 per cent in the Yakutsk area of Eastern Siberia.

European Russia, where the main centres of consumption and the export terminals are concentrated, has the remaining 22 per cent of USSR known reserves.

It will obviously be costly to develop the Siberian reserves, where the largest gas accumulation is in the Tyumen region of Western Siberia. The largest of the major fields here is the Urengoy, which has reserves that are more than three times as large as those of the Dutch Groningen field.

An indication of the scale of the problem of exploiting the major Siberian natural gas reserves is that they lie some 1,500 miles from Moscow and Leningrad, 2,000 miles from the Black Sea and almost 3,000 miles from the Pacific coast.

Clearly, the USSR has vast natural gas resources which will no doubt prove a valuable asset to the country for a long time to come but it is unlikely that they will make a significant contribution to the world's medium term energy problems.

America's natural gas reserves will certainly be needed for internal use and, indeed, will need to be supplemented from outside (as the crisis caused by gas shortages in the winter of 1976/7 has shown). The 1972 Report by the National Petroleum Council put ultimate discoverable reserves at 1,857 trillion cubic feet – a figure which is best comprehended by referring to its oil equivalent of about 45 thousand million tonnes (compare the *National Energy Outlook February 1976* figure of oil measured reserves of 4,700 million tonnes, at 1973 prices).

This figure represented recoverable gas and not "gas-in-place" and was about 7½ times the published proved reserves. Little more than a third of this gas had been actually found by January 1st 1971. Over 40 per cent of the gas which remained to be discovered was thought to lie in Alaska or offshore in deep water and therefore costly to recover.

There is, however, very great uncertainty about the size of the undiscovered natural gas resources and in more recent times the US Geological Survey has cut its estimates heavily. In the Federal Energy

Administration's *National Energy Outlook February 1976,* already quoted in connection with US oil reserves and resources, a mean estimate of total recoverable resources, based on US Geological Survey data, gives a figure of 961 trillion cubic feet. This comprises measured reserves of 237 trillion cubic feet, inferred reserves of 202 trillion cubic feet and 522 trillion cubic feet undiscovered recoverable resources.

I have made these separate references to the natural gas reserves of the USA and USSR because of their size and their importance to the world energy economy. There are reserves in many other areas, including the Middle East and North Africa and the North Sea, which I shall discuss in detail later.

Coal in Abundance

Now we approach what can be regarded as the crucial factor in the medium and long-term energy situation – the wealth of coal which in recent times has been to a large extent neglected.

There are vast reserves of coal world-wide and this abundant energy source is probably more widely distributed than any other form. It is particularly significant that many of the industrialised countries, where energy needs are high, are better supplied with indigenous coal than other fuels. The only major regions which are relatively badly off for coal, according to the World Energy Conference, are Latin America and Africa. Since these also happen to be among the least thoroughly explored regions, much may yet be discovered there; and indeed quite large reserves are being found in certain Latin American countries such as Venezuela, Colombia, Mexico and Brazil.

According to the latest estimates, technically and economically recoverable solid fuel reserves should suffice for over 200 years' production at present rates. This, however, is a poor guide since, as and when the need arises, there are far greater quantities in the ground which may not be recoverable under present circumstances but could easily become so as the world's other energy resources dwindle.

For example, according to estimates given to the 1977 World Energy Conference, total world geological resources of solid fuel are over fifteen times as high as the technically and economically recoverable reserves, while, as mentioned in the previous chapter, Paul Averitt in a 1967 United States Geological Survey Bulletin, has suggested an even higher figure.

Contenting ourselves with the figure for technically and economically recoverable reserves of solid fuel (that is coal and lignite – in terms of energy content) we find that Western Europe has about 14 per cent of the total, the USA 28 per cent, Asia 22 per cent the Soviet Union 17 per cent and Eastern Europe 6 per cent, leaving 13 per cent for the rest of the world (Australia, South Africa and Canada having nearly 80 per cent of this). These figures may over-estimate the shares of the more highly developed areas where exploration has reached a high level.

It certainly under-states the eventual resources of the Soviet Union which, on this basis of geological resources, has nearly half of world reserves. The figures for Asia relate largely to the Chinese Peoples' Republic about whose coal reserves, as of their other forms of energy, relatively little is known so that these figures are even more speculative than most.

Even in Western Europe, coal resources are available in the ground to support much higher production rates than those obtaining at present. Having said that, however, there remains the problem of extracting coal from fields where geological conditions tend to be more difficult than in many other areas.

After the United Kingdom (I shall be dealing separately with our coal industry), the Federal Republic of Germany is Western Europe's largest coal producing country. In the World Energy Conference Survey for 1968, Germany's measured reserves were put at 70,000 million tonnes with a further 160,000 million tonnes at depths between 3,940 and 6,500 ft. – too deep for mining under the economic conditions of the time. Since that date, a large number of mines have been closed, so that the latest estimate of technically and economically recoverable reserves of hard coal is about 24,000 million tonnes. Total geological resources are however now put at 230,000 million tonnes. There are also 16,500 million tonnes c.e. of brown coal (lignite) of which 10,500 million tonnes c.e. were reckoned to be technically and economically recoverable.

Deposits of coal and lignite are widely distributed throughout France but only the Nord/Pas de Calais and Lorraine basins are really of more than local significance. The latest estimates set solid fuel reserves, technically and economically recoverable, at 438 million tonnes c.e. and geological resources at about 2,400 million tonnes c.e. At present extraction rates, the recoverable reserves would last for no more than about 20 years, but surveys in 1968 and previously give much higher figures.

French coal is expensive to mine (especially in the Nord/Pas de Calais coalfield) and this is also true of Belgium whose output has been drastically cut in recent years. Spain is another coal producing country with reserves which are relatively costly to exploit.

Outside the USA and the USSR, China is believed to possess much the biggest solid fuel resources of all, though little reliable information is available. Besides Western Europe, there are significant solid fuel deposits in many other areas, such as Eastern Europe, India and Australia, to mention only a few whose reserves are not only important for their own internal economies but either are, or could be, important in the international trade in coal, as I shall explain later.

Coal Giants

Already in this chapter I have referred to the important role of the USA and USSR in world energy matters and this applies also to coal. Between them these two countries have around 45 per cent of the world's technically and economically recoverable reserves of coal, and nearly three quarters of the total geological resources according to estimates given to the 1977 World Energy Conference.

Paul Averitt's estimates in the US Geological Survey paper already quoted in connection with oil reserves, put the United States total coal-in-place at 2,900 thousand million tonnes of which 136 thousand million tonnes of recoverable coal were located in seams of comparable thickness and depths to those being economically mined at the time. About 40 thousand million tonnes consisted of coal which could be surface mined (equivalent to British open-cast mined coal). The latest estimates given to the World Energy Conference for technically and economically recoverable hard coal (113 thousand million tonnes) are rather below Averitt's recoverable coal figure but there are also 64 thousand million tonnes coal equivalent of subbituminous coal and lignite shown.

With technically and economically recoverable reserves of 110 thousand million tonnes c.e. and geological resources of 4,860 thousand million tonnes c.e., the Soviet Union is the other coal giant on the international scene. Paul Averitt's estimates put the ultimate possible total at an enormous 8,620 thousand million tonnes, or well over half the world stock!

Over 90 per cent of these huge resources are reported to be in Asiatic Russia where, for example, the Kuznetsk coalfield in Western Siberia has some 900 thousand million tonnes of reserves, of which some 30 per cent is of coking quality. To transport this coal to European Russia requires journeys averaging over 900 miles, with some journeys up to 1,860 miles. In Eastern Siberia, the Tungus, Kansk-Achinsk, Irkutsk and Minusinsk basins are claimed to have 3,700 thousand million tonnes of coal. The Far Eastern region of Russia has 25 thousand million tonnes in the Bureya basin and 20 thousand million tonnes in the Sakhalin basin.

In the Soviet Union, as elsewhere in the world, the coal reserves are obviously available in the ground for almost indefinite exploitation, if we discount the problems of extraction.

This is the problem which faces coal industries the world over, including our own country to a certain extent as we shall see. The longer preparations are delayed to exploit these reserves the more difficult will the position be when the reserves are urgently required.

How Britain Stands

My next task is to set Britain's reserves of the three major sources of energy – coal, natural gas and oil – against the wider world distribution of these fuels. Within the last decade or so, the situation has changed dramatically for the better in the short term and, I believe, in the mid- and long-term also, but only if we adopt the right priorities and policies.

Until 1965, it would have been said that Britain's only indigenous energy source was coal, for the sporadic search for oil and natural gas on land had been largely unsuccessful and hydro-electric potentiality was limited by reason of geographical factors.

This was how matters stood when natural gas was found in the southern basin of the North Sea in 1965, the first supplies coming ashore in 1967. Six fields are now being exploited – the "big five", West Sole, Leman Bank, Hewett, Indefatigable and Viking, and the smaller Rough field – delivering 39.4 billion cubic metres in 1976, or 98 per cent of total UK gas consumption. The bulk of the balance was imported in liquefied form from Algeria.

Gas is also under contract from the UK/Norwegian Frigg field (now coming ashore) and the Brent field (expected in 1980 or soon after). Frigg gas, within a couple of years, should build up to about 15 billion cubic metres a year while Brent will add a further 5 billion cubic metres a year.

Algerian natural gas, at about 1 billion cubic metres a year, will continue to arrive until the present contract expires in 1980.

Further possible supplies could come from the gas which is extracted with oil (known as "associated gas") but which is normally flared off at the production platform. This gas is not easy to collect and bring ashore, mainly because it is produced in relatively small volumes at each individual production platform, but the Brent contract covers associated gas and some is also being obtained from the Forties field.

An ambitious scheme has been proposed to construct what would in effect be a ring main to collect associated gas from a number of oil-fields in the northern waters of the North Sea. A study company – Gas Gathering Pipelines (North Sea) Ltd. – has been set up, with the British National Oil Corporation and British Gas as the main participants, to establish the economic, technical and marketing viability of such a scheme which, it is reckoned, could supply between 10 and 15 billion cubic metres of gas a year for about 12 years, or about a third of present total UK gas consumption.

In addition to the proven gas reserves, exploration for gas is

1 *(top)* Parkside Colliery, Lancashire, opened in 1964

2 *(centre)* The Longannet complex, Scotland, showing the main conveyor road

3 The Author underground at Seafield Colliery, Scotland, April 1977

4 Coal reclaimer in operation at the NCB's Immingham terminal, the largest coal exporting facility in Western Europe

5 *(top right)* Laboratory rig of refinery for conversion of coal to oil at the NCB's Coal Research Establishment

6 *(right)* Work in progres on the Selby coalfield development, the world's largest deep-mining project

7 Testing a powered roo support at the NCB's Mining Research and Development Establishment

8 *(top)* The remains of the former Parkhouse Colliery, Derbyshire, before opencast working, showing the 150 foot high tip containing over one million cubic yards of material

9 *(bottom)* The site restored after extraction of 300,000 tons of high quality steam coal

continuing, though on a lower scale than oil prospecting. It is generally reckoned that further major reserves are unlikely to be found in the southern basin but that discoveries of gas, either on its own or in association with oil, are possible in other parts of the UK continental shelf.

How does all this leave Britain's natural gas reserves as we enter the last quarter of the present century?

Estimates of UK oil and gas resources are given each year in a Report *Development of the Oil and Gas Resources of the United Kingdom*, presented to Parliament by the Secretary of State for Energy. It is familiarly known as the Brown Book.

According to the 1977 Brown Book, proven gas reserves remaining at the end of 1976 amount to 809 billion cubic metres. This total includes the reserves both in the southern and northern basins of the North Sea and also finds in Liverpool Bay; it covers fields under production or under contract to British Gas and not yet producing as well as other discoveries (including associated gas) but not yet contracted to British Gas. Total southern basin proven reserves are reckoned to be 513 billion cubic metres.

Northern basin* proven reserves amount in total to 297 billion cubic metres. The total of these two sectors, or 809 billion cubic metres, as already noted, is slightly less (6 billion cubic metres) than the previous end-year figure, a reduction of 39 billion cubic metres in the southern basin having been largely offset by an increase in the northern basin. The latter increase is however in gas associated with oil discoveries, and gathering this gas can pose problems, as I have said.

The non-proven (probable and possible) reserves estimated to exist in known discoveries – in addition to the proven reserves – are mainly in the northern basin. Out of total "probable" reserves, 113 billion cubic metres are expected from the southern basin and 158 billion cubic metres from the northern basin, making probable finds of 272 billion cubic metres in all.

"Possible" reserves are reckoned at 362 billion cubic metres, of which 65 billion cubic metres are thought to be located in the southern basin and 297 billion cubic metres in the northern basin and elsewhere.

Putting all this together, the Brown Book gives a combined total of proven, probable and possible offshore gas reserves in known discoveries of 1,443 billion cubic metres. On a simple arithmetical basis this is 37 years supply at the current annual rate of UK consumption – proven reserves would represent 21 years supply – but this of course is not a valid indication of how long the gas will actually last.

It is unlikely that gas consumption will remain at the recent level and certainly British Gas policy looks towards considerably higher

*with Liverpool Bay.

consumption levels. Existing discoveries, it is suggested, should support annual extraction rates of over 60 billion cubic metres (about 90 mtce) by the early 1980's and maintain this level until well into the decade. Any further discoveries and contracts will extend this period.

Many factors will determine the date by which natural gas supplies will reach the inevitable production plateau, after which they will no longer be able to match rising demand, until eventually an actual decline of supply begins.

On present evidence it would be reasonable to assume that the plateau will be reached before the end of the century, even though supplies at a slowly diminishing rate may still be coming in until well into the next century.

Britain's Oil

Much has been said and written about North Sea oil since the first discovery in the British sector in 1969 began what is obviously going to be a major contribution to our energy problems from now on. It is not my purpose here to add to the mounting mass of facts and figures, opinions and speculations on Britain's North Sea riches but to consider this natural wealth in the context of the long-term energy problem and the theme of this book.

According to the Brown Book, total proven reserves at the end of 1976, in the UK sector of the North Sea, amounted to 1,380 million tonnes. This is much the same as a year earlier, indicating that the pace of discovery has slowed down, at least temporarily. However, the *Oil and Gas Journal* gives a much higher figure for proven reserves – about 2,300 million tonnes (the sum of proven and probable reserves in the Brown Book) – another illustration of the margin of doubt in any figures of oil reserves.

When probable and possible reserves are included, the Brown Book total reaches 3,200 million tonnes. These figures relate to areas of the UK continental shelf which are already licensed and when account is taken of areas which have been designated for further future exploration but not yet licensed, and areas expected to fall to Britain when dividing lines are agreed between Britain and other countries, such as Ireland, France and Norway, total reserves could possibly reach 4,500 million tonnes.

Fourteen commercial fields are listed in the Brown Book as being in production or under development – Argyll, Auk, Beryl, Brent, Forties, Montrose and Piper which were in production by the end of 1976 plus Claymore, Cormorant, Dunlin, Heather, Ninian, the UK part of the Stratfjord field (which straddles the British-Norwegian median line) and Thistle. The first oil came ashore in June 1975 from the Argyll field, followed by the Forties field (probably the biggest in the UK sector) in November of the same year.

By the end of 1975, these two fields had delivered 1.1 million tonnes. With five more fields coming on stream during 1976 production rose rapidly and when the year ended, oil was flowing at a rate of almost 2 million tonnes a month. Total production during the year amounted to 12 million tonnes. It has continued to build up during 1977 and the anticipation is that as other fields come on stream, and as the existing seven fields increase production, output will reach between 90 and 110 million tonnes in 1980.

During the early 1980's, production in the range 100-150 million tonnes is expected and it is the hope of output on this scale which has led to the anticipation of self-sufficiency in oil by 1980. This does not mean that we shall obtain all our oil from the North Sea as we shall still need to import some heavier, relatively cheaper crudes to give us the "mix" we need to meet our requirements for the various types of petroleum products.

We should, however, have some high quality, premium value North Sea oil surplus to our needs for that type of product and this we shall be able to export, either in crude or refined form.

The figure for proved reserves given in the Brown Book would only support an output of 100 million tonnes a year up to 1985, after which output would begin to decline. If total recoverable reserves were to reach 4,500 million tonnes, output might build up to 200 million tonnes a year, or more, by 1985 but it would start to fall off before 1990. If it were restricted to around 150 million tonnes a year from say, 1982, it would probably be able to support that level until about 1995. Even in that case, output would be falling quite fast by the end of the century – at a time when energy shortages could be developing on a world-wide basis.

Built on Coal

Recoverable reserves of British coal are probably some 45 thousand million tons – several times greater than the coal equivalent of the ultimate combined resources of both North Sea oil and gas. At present rates of production these reserves would last for well over 300 years. The total coal-in-place is estimated at 190 thousand million tons, but inevitably not all will be workable.

There has been confusion in the past because of wide variations in quoted coal 'reserves', reflecting radical changes in proposals for the industry. In 1905 a Royal Commission assessed British coal reserves at some 142 thousand million tons. Allowing for subsequent discoveries and the coal which has been extracted since then the corresponding figure would now be about 200 thousand million tons. At that time there was no feasible substitute for coal as a source of bulk energy supply. Later assessments have allowed for coal which has become sterilized by mining operations, or has become uneconomic. For example a 1946 Report by the Coal Survey Organisation of the Department of Scientific and Industrial Research estimated the UK total at about 35.5 thousand million tons, of which they expected some 20.5 thousand million tons to be obtainable over the next hundred years.

During the 1960s however with increasing availability of oil at low prices, there was thought to be no case for new coal capacity for the foreseeable future. We concentrated on reserves which were available or might become available only in existing collieries or those actually planned. Also mechanised mining, essential to contain the competition from oil, reduced the percentage of reserves extracted. Consequently the estimates of reserves fell from 14.6 thousand million tons to 6.4 thousand million tons between 1965 and 1969.

Large tonnages of reserves were discarded as being no longer economically workable. These lower figures, however, were directly related to the low priority given to coal during that period and bore no relation to the reserves actually in the ground. This was made clear at the time of the 1974 World Energy Conference when the NCB on the knowledge then available, reported total ultimate resources in the UK as about 160 thousand million tonnes of which 97 thousand million tonnes were in known deposits and the remainder classified as "indicated and inferred."

We now use the term "coal-in-place" (rather than "ultimate resources") to indicate the coal existing in the ground. The estimate of 190 thousand million tons given above does not include coal below 4,000 feet or under the North Sea which would be unworkable with present technology and economics.

That part of the coal in place (45 thousand million tons) which could be recovered using established technology we term "recoverable reserves". We have introduced the category "operating reserves" to indicate the part of recoverable reserves which are fully proved as suitable for mining, and which are either accessible to existing collieries and opencast sites or available for identified new collieries or opencast sites. The operating reserves at existing mines are currently put at some 4 billion tons and a further 2 billion tons have been identified for new mines. There will be a progressive upgrading of recoverable reserves into operating reserves. The intention is to maintain a "stock" of at least 50 years of operating reserves at any one time. This is provided by the current 6 billion tons.

Particularly well-known now is the big discovery in the Selby district, a major eastwards extension of the Yorkshire field, where some 450 million tonnes of reserves are to be developed at a proposed extraction rate of 10 million tonnes a year. Even bigger is the reserve in the North-East Leicestershire area where 500 million tonnes of reserves are known to exist.

Even more recently, over 200 million tonnes of workable coal has been discovered in an area west of Coventry and north of Kenilworth. The coal in all three areas is at depths between 1,000 and 2,500 ft.

This by no means completes the tale of success. There are reserves of 130 million tonnes at Park, in Staffordshire, which could support an annual output of at least 2 million tonnes. And other areas where there are good prospects for finding additional reserves include Musselburgh (on the southern shore of the Firth of Forth), and various locations in Yorkshire, the Midlands and elsewhere.

It is sometimes assumed that, because we have been mining coal in Britain for centuries, all the best coal must have been worked and that all these new reserves will be less productive. The evidence of recent exploration suggests otherwise.

What we have to consider when evaluating reserves are the three major parameters of seam thickness, the amount of disturbance below ground (which results in discontinuinty of the seams) and the depth at which the seams are found.

In terms of seam thickness, the classified reserves at existing collieries will enable the average working section to be broadly maintained, at least to the year 2000, but this is being added to considerably by the new reserves being identified by the extensive exploration programme.

Very large potential reserves exist in areas of low geological disturbance, especially in the extensions of the productive Yorkshire and Midlands fields. The working depth of a major part of the potential reserves, although greater than experienced in typical nineteenth century mining, is no different from conditions experienced in the most productive of modern collieries.

So far, I have referred only to coal reserves below the ground on which we live. Additionally there is coal in abundance under the North Sea, where much of the oil and gas now being exploited lies above coal deposits. In fact, some oil exploration boreholes are reported to have found coal seams up to 50 ft thick!

Though coal is already being extensively mined under the sea from shore-based collieries, it will, admittedly, be a long time before the technology is developed to extract coal far out on the continental shelf. There is little doubt, however, that it will eventually be commercially possible to develop these resources long before the known shore-based reserves are exhausted.

Most of the reserves on land are of deep-mined coal, but there are also reserves sufficiently near the surface for extraction by open-cast methods which have produced about 200 million tonnes over the past

33 years, and which at present yield about 11 million tonnes a year, or about 9 per cent of total output.

Coal in the ground, however, is not fuel in the power station, the steel works or the home. The reserves have to be exploited and this is a question I shall deal with in detail later.

Moving Energy Around

Before we discuss the exploitation of our energy reserves I want to return to the international scene and, to consider the movement of energy around the world, from areas where reserves are abundant to others where they are not. Clearly, it is no good giving optimistic reports of huge energy reserves if they are unlikely to reach the consumer. The problem is compounded by the fact that some of the biggest energy consumers have the smallest indigenous resources. It has thus become increasingly necessary to transport the main forms of primary energy – coal, oil and natural gas – in ever larger quantities and over longer distances.

At first glance, it might appear difficult to transport a commodity as bulky and heavy as coal in large quantities and even more difficult to move the vast volumes of petroleum products which the world requires. Yet, in fact, there has been trade in oil for many years and, as long ago as 1913, British coal exports amounted to as much as 73 million tonnes. The first oil tanker was built on Tyneside in 1886 and, surprising as it may seem, the "super-tanker" of 100,000 tonnes d.w. plus has been on the seas for nearly twenty years.

Natural gas poses more difficult problems, because of its vast bulk in relation to heating content in its gaseous state, but, when liquefied, large quantities can be carried in special cryogenic tankers and gasified at the point of use. Developments in high pressure steel pipeline technology enables economic quantities of gas to be transported over great distances, on land and beneath the sea, although sub-sea pipelines are very expensive.

How the Energy Trade has Developed

It is first necessary to trace the rise of world trade in energy because, as I have said, although both coal and oil have been exported for many years, until 1925 most countries made do with their indigenous supplies, including the United States and Western Europe, and the trade in fuels was mostly between countries in the same general area. Only Mexico produced fuel mainly for export and her oil exports of 13 million tonnes a year were greater than the combined total of all other countries producing mainly for export.

By 1950, the trading position had changed markedly. Western Europe, including Britain, was no longer a net exporter of coal and the region, as a whole, had quadrupled its oil imports, becoming a net importer in respect of about a sixth of total energy requirements. The United States was still exporting coal, mainly to Canada, but, in spite of her own large oil resources, was now importing over 30 million tonnes of oil a year, or more than 10 per cent of her usage.

Mexico had cancelled her oil concessions to overseas countries by 1950 and was no longer a substantial exporter while Venezuela and the Middle East had emerged as major oil exporters, each producing between 80 and 100 million tonnes, almost all of which was exported. The significant fact had emerged, even at that time, that these two regions alone were accounting for 30 per cent of the world's production of oil, or about 8 per cent of total world energy production.

During this period, the Soviet Union and Eastern Europe had almost trebled their share of world energy production, but although they were emerging as net exporters, the amounts exported remained relatively small. Japan, another influence of growing importance on the world energy scene, was still largely dependent on coal in 1950, being almost self-sufficient at that time.

For the next few years, the trend continued. Exports of Middle East oil continued to grow and Venezuelan production also increased substantially. The importing countries became more and more hungry for energy, particularly oil, and, by 1956, Western Europe depended on imported oil for a quarter of its total energy requirements, although Britain at that time only imported around one-eighth of its energy needs. The United States almost doubled its oil imports between 1950 and 1956, reaching 59 million tonnes a year by that date. Japan, however, was still only a small importer, with 13 million tonnes in 1956.

Oil imports continued to grow after the Suez crisis of 1956, especially in Western Europe and Japan. Britain's net oil imports rose more than threefold between 1957 and 1972 (from 33 million to 107 million tonnes) while EEC imports (that is, the original "Six") increased over the same period from 61 million to 419 million tonnes. In 1974 the enlarged EEC imported 560 million tonnes.

Still more significant was the trend of imports into the huge US market and Japan, both of which greatly increased their demands on the world oil stock during the period 1957-74. United States imports rose from 60 million to 292 million tonnes while Japan recorded a staggering rise from a mere 18 million to 256 million tonnes in 1974.

Taking these figures together, with other smaller imports in other parts of the world, including South America, total net imports of oil by the importing areas of the world increased from under 250 million tonnes in 1957 to over 1,400 million tonnes in 1974.

Export Sources

So far, I have looked at the problem mainly from the import side. A closer look at the world's major net exporters of crude oil shows that in 1974 they produced about 1,540 million tonnes, of which the Arab states accounted for almost 900 million tonnes, or 58 per cent. The other principal countries producing for export were Iran (302 million tonnes or 20 per cent), Venezuela (159 million tonnes or 10½ per cent), Nigeria (112 million tonnes or 7 per cent) and Indonesia (68 million tonnes or 4½ per cent).

These figures underline once more the strong bargaining position of the OPEC countries. Compared with these countries, the world's other major exporters, the Soviet Union and Canada, made only a relatively minor contribution to world imports in 1974. The Soviet Union had net exports of 76 million tonnes of crude oil and 31 million tonnes of refined products, the greater part of which went to Eastern European countries. Canada exported over half its output of 93 million tonnes of crude, because it is more economic to import oil into its eastern states, due to their great distance from the indigenous sources, but Canada is only a minor net exporter.

In their varying ways, the United States, the Soviet Union and Japan have such a big influence on the world oil situation that it is worth giving special consideration to these three areas.

Although, as has already been seen, the United States had the world's highest oil production rate until two years ago, output has tended to decline since 1972 while consumption has risen further, so that there has been a big increase in imports. In 1972, Venezuela and Canada provided over half the United States imports of crude oil, but their share has fallen off greatly since then. In 1976 Arab states accounted for nearly 45 per cent of US crude imports, of which Saudi Arabia provided half. The other main OPEC members (Nigeria, Indonesia, Iran and Venezuela in that order) accounted for 40 per cent and Canada for 7 per cent.

Fuel consumption in the Soviet Union and Eastern Europe has increased more rapidly than in most other parts of the world, but production has kept pace. As a result, the area as a whole is not only self-sufficient in energy but has continued to export both oil and coal to a limited extent, though up to now it is a net importer of natural gas.

So far as oil alone is concerned, the USSR is now the world's largest producer. Unlike the other major oil exporters, the Soviet Union can reach many of its markets by overland pipeline and the biggest Western Siberia oilfield, the Samotlar, is connected by a 48 inch pipeline, first to the Soviet Ural-Volga area and thence by a connecting pipeline system to Poland, West Germany, Czechoslovakia and Hungary.

Net Soviet oil exports in 1974 were rather over 100 million tonnes, the greater part of which went to the other members of the Eastern Bloc. Most of the remainder is sold to Western Europe, but Japan and Cuba also get sizeable amounts.

Japan, with hardly any indigenous oil, must inevitably have a continuing effect on the supply position of the world as a whole. The country's prodigious expansion in economic power has had an overwhelming effect on her energy requirements, particularly oil, imports of which increased from less than 2 million tonnes in 1950 to over 250 million tonnes in 1974.

Most of Japan's petroleum is imported in crude form, the principal suppliers, in 1976, being Saudi Arabia, Iran, Indonesia and United Arab Eminates. In contrast to Britain and Western Europe generally, Japan had been getting about half her imports from non-Arab states (mainly Iran and Indonesia), but the proportion has now fallen to one third.

Natural Gas Exporters

Unlike both oil and coal, which have been moved around the world for some considerable time, there was very little world movement of natural gas during the 1950's and 1960's. In 1969 more than three-quarters of such trade as was taking place in natural gas consisted of sales from the Canadian fields to the United States and vice versa and from the Dutch Groningen field to other European countries. These movements between neighbours had increased some 250 per cent by 1974, mainly owing to the growth in Dutch supplies. They were however a smaller proportion (64 per cent) of total exports mainly because the USSR and Iran had become sizeable exporters. Despite this very large increase less than 9 per cent of total world production of natural gas in 1974 crossed any frontier.

The Soviet Union has a two way traffic in natural gas and for a few years was a net importer, taking more from Iran and Afghanistan than she exported. But she has now stepped up her exports and they are being expanded further. Most of her prolific gas fields are in Western Siberia which, however, is not only very remote from the Soviet Union's own centres of demand but is even further away from export markets. Considerable supplies are already distributed by pipeline to the East European countries and to West Germany, Italy, Austria and Switzerland. Plans are under discussion for further exports, in liquefied or gaseous form, to Japan and the United States.

The Coal Trade

Perhaps the most surprising feature of the development of international trade in the past decade or so has been the relatively small part played by coal – the fuel which is demonstrably in most abundant supply and which, with the evolution of bulk sea transport, allied to loading and unloading techniques, can be moved over considerable distances at reasonable cost.

According to 1974 figures, total world trade in coal, in relation to total world production, was just about the same as for natural gas, namely, rather less than 9 per cent. The three largest exporters are the United States, Eastern Europe (especially Poland) and Australia with South Africa beginning to play an important role. Western Germany exports sizeable amounts to certain other Community Countries, especially France, Italy, Belgium and Holland.

In 1974, the United States had net exports of 53 million tonnes, nearly half of which went to Japan, with Canada and Western Europe sharing most of the other half.

Eastern European countries were net exporters of coal in 1974, to the tune of 44 million tonnes, more than half from Poland, which had doubled solid fuel production since the early 1950's and is now the world's fourth largest producer, behind the USA, USSR and China. Most of Poland's coal exports go to Western Europe, apart from the supplies sent to her neighbours in the Eastern Bloc (including Russia). Russia mainly exports to Eastern Bloc countries but sent 8 million tonnes to Western Europe in 1974.

Australia exported 30 million tonnes in 1974, a six-fold increase in ten years while Canadian exports have also been growing fast. However, Canadian exports only stood at 11 million tonnes in 1974 and she imported rather more coal, due to the economics of transporting indigenous coal over great distances, compared with the movement of American coal, cheaper to produce, over shorter distances.

Both the United States and Poland export coal to Western Europe. In 1974, the other 8 members of the European Community were net importers of 32½ million tonnes of solid fuel, of which 14½ million tons came from Poland and 9 million tonnes from the Soviet Union, and three million tonnes from Australia. We in the UK sent 2 million tonnes to the Community.

It may seem almost unbelievable that coal should be brought half way round the world to a Continent which still has considerable reserves. This underlines the unfortunate effects of the rundown of indigenous European coal production in the post-war period. It is also in part due to relatively low production costs in some other countries (especially where opencast mining plays a significant role) and partly to the need for particular qualities (e.g. coking coal and anthracite).

As world dependence on coal increases, so too will the international trade in coal. Coal mined competitively in modern conditions should have good export opportunities. The latest estimates, made at the World Energy Conference in Istanbul in 1977, showed that 500 million tonnes of coal could be moved internationally by the year 2000 compared with some 200 million at present.

Exploiting the Reserves 7

When one sets down the world's energy reserves, first on a global basis and then by major regions, as we have done in the two previous chapters, there is a danger that such apparently huge figures, especially those which relate to possible and as yet undiscovered resources, will breed complacency. It is only necessary to make two points, however, and all unjustified optimism should disappear.

First, as we saw in Chapter 4 (which dealt with demand) the world is using energy at a great rate, even with the slower growth in consumption which has followed the 1973 crisis, and it is virtually certain that the rate of usage will continue to grow.

Even if real prices rise to double their present level, the expectation is that world energy demand, from 8,600 mtce in 1974, will increase to around 12,500 mtce by 1985 and a staggering 20,000 mtce by the year 2000.

These are totals for the whole of the world. Restricting ourselves, for the reasons I gave in Chapter 4, to the non-Communist countries and with whose problems this book is chiefly concerned, the forecasts for this sector, as we have seen, are also formidable. From about 6,150 mtce in 1974 the expectation is that demand should reach 9,000 mtce by 1985 and 14,000 mtce by 2000. By the year 2000 the world is unlikely to be able to pick and choose between different fuels, but will need to develop all economic sources to the full. For most major industrialised areas of the non-Communist world likely demand for energy will exceed what they will be able to produce from their own indigenous resources and they will have to rely substantially on imports, especially of oil but increasingly also of coal.

Secondly, and still more importantly, the reserves listed in Chapters 5 and 6 are still in the ground, whence they have yet to be extracted and their exploitation will in many cases become progressively more difficult and therefore more expensive as the more easily worked or more accessible reserves are exhausted. In this chapter we look at the problems of exploiting these reserves. Will the world be able to exploit them to the extent required, and in time, to meet that 2000 objective?

It is a many-sided problem, compounded of geology, economics, politics and technology, with close inter-relations between all the trends under these headings, as a few of the questions we must ask ourselves will soon reveal. What technological advances will be necessary to wrest the resources from the earth and beneath the sea in ever more inhospitable locations? What will be the real cost of obtaining the reserves and will it be possible to develop them at prices which will find buyers? These are some of the economic/technical problems but they cannot really be separated from political and socio-political questions. They range from the policies which will be adopted by countries with surplus energy for export (but perhaps with reasons for not wishing to do so) to the increasing pressures from environmentalists.

All these problems will affect every country, including our own, where we in the coal industry are not only acutely aware of what must be done but are taking steps to lay down the necessary long term plans. Determined action, if undertaken world-wide by all fuel industries, particularly if equally strong action is taken to ensure energy conservation, would increase output of many of the fuels beyond the levels expected at present and so hold down the rise in energy prices which would otherwise occur.

Although this chapter will present the future fuel picture as it appears on present prospects, we can keep in the back of our minds the encouraging thought that the world can make things much easier by taking stronger action than may seem likely at the moment.

Key areas in the non-Communist world obviously include Western Europe, which has some indigenous energy to develop, but must remain one of the biggest importing areas, Japan, which will perhaps remain more dependent on imported energy than any other major power, the United States, with important reserves to exploit but still likely to make demands on the world energy stock, developing countries in Asia, Africa and Latin America which will require more energy as they advance in technology and industrialisation, and the OPEC countries, with their dominant role in the exploitation of the world's biggest known reserves of oil and natural gas for export to the rest of the non-Communist world.

I shall be dealing particularly with some of these areas, and separately with Britain which occupies an unusual position at the present time in that we shall soon be self-sufficient in energy, with the possibility (for a period) of becoming a net exporter. This, however, does not give us the opportunity to opt out of the world energy race – on the contrary, all the appearances are that we could be coming back on to the world import market, when our indigenous reserves of oil and gas are declining, just at the time when the world supply situation is becoming tighter and prices are soaring.

Britain is therefore very much a part of the world scene, which I shall now consider, starting with Western Europe, since this is the region with which we are bound to be most closely concerned.

West European Prospects

All the evidence suggests that the EEC (less the UK) will remain heavily dependent on imports from outside the area, in spite of the objectives which were set back in 1973, intended at that time to reduce dependence on imported energy and to improve the security of such supplies as will still have to be imported.

There is every incentive, therefore, for EEC countries to seek to exploit the indigenous reserves they possess, and in particular to make full use of their coal. Note, however, that in 1955 the countries of the EEC (less the UK) produced 280 mtce of solid fuel; since then the figure has fallen by over 40 per cent. It might have been thought that the logic of the situation would have encouraged a revival of coal production in the rest of the EEC, in contrast to the pre-1973 expectation that it would continue to contract.

Unfortunately, the problems of reviving coal output on the Continent are serious. Their coal has always tended to be less competitive than in most other coal producing areas, for geological and cost reasons. Although there is still a great deal of coal "in the ground" in several Continental countries, the Governments concerned do not yet seem to believe that there is an economic case for expanding coal output and for giving appropriate encouragement to producers.

If I may offer a personal opinion, I feel that the situation has been mis-read. There is much to be gained by raising the output of any coal which can be expected to compete with oil at the higher prices likely in the future, although this will mean the commitment of considerable financial resources to replace collieries closed for economic reasons or through exhaustion of resources. It will take time to make the necessary investment even to stabilise production at present levels, let alone raise it, but I believe that additional coal could be produced in the EEC countries at prices which would be competitive with alternative fuels, especially on the assumptions of a doubling of real prices by the end of the century (it has even been suggested that this might occur a decade earlier).

Time is growing short and there are no signs as yet of a reversal of the decline in coal production which, in the rest of the EEC, has fallen still further since 1973. On present prospects, production could be even lower by 2000 when net coal imports will have grown appreciably above their 1976 level of about 40 million tons.

A considerable proportion of the imports will consist of coking coal, since steel industry coke ovens take a much larger proportion of total coal supplies than in Britain; in addition sizeable quantities of steam coal for power stations are likely to be required, particularly if EEC Governments follow the Commission's policy of encouraging the use of coal, rather than oil or natural gas, at power stations.

Natural gas production in Continental EEC countries has risen from almost nothing in 1950 to 170 mtce in 1974 (and a little higher still in 1976) or more than 40 per cent of the total indigenous energy production of the area. Supplies are unevenly distributed, however, with the Dutch reserves, mainly in the vast Slochteren field, forming nearly four-fifths of total resources. As this field is approaching its peak, the Dutch Government has adopted a Gas Marketing Plan, under which priority is to be given to maintaining supplies to residential and small industrial consumers. The Plan states that, after 1978, sales to power stations, large industrial users and to export markets will be gradually phased out, at a rate which will depend on the discovery of additional reserves.

Natural gas production in Continental EEC is expected to have turned down before 1985 and to be substantially lower by 2000. Since consumption estimates show a considerable increase over today's levels, the gap which will need to be closed by imports will be a big one – perhaps 100 mtce by 1985 and double as much by 2000. Some of the imported gas will be supplied by Norway (the one sizeable producer in Western Europe which is outside the EEC). This is likely to be only a fairly small proportion of total needs, since Norwegian gas is also being supplied to Britain, for example from the Frigg field. Some gas might also be supplied from Britain if a pipeline is constructed across the Channel.

Oil production in the Community (apart from Britain) was only 16 mtce in 1974 and there are no firm prospects of a significant increase. Here is the really big problem, with the oil import requirement likely to continue to build up from 655 mtce in 1974 – though it has since dropped slightly – to approaching 1000 mtce by the end of this century, always assuming that supplies of this order are available.

This estimate of oil requirement assumes that nuclear power production in the Community outside Britain will reach about 800 mtce by 2000. If not, still more oil may be needed as a replacement for the shortfall. The figure of 800 mtce is not unreasonable in the light of the plans of the countries concerned – although these plans do not look as far ahead as 2000 – but, even so, it may prove unattainable.

I base these doubts on a number of factors, including the environmental objections which will have to be overcome, the time scale problem, the physical resources available, the cost of such a programme and whether electricity demand will grow fast enough to support a programme on the scale envisaged.

Each of these factors could be discussed at great length but all I propose to say here is that environmental objections to nuclear power are growing more powerful, that the time needed to design, build and commission existing reactor types (let alone the new systems which will be needed) may make it physically impossible to increase installed capacity in the scale required and that the cost may be prohibitive, given that other energy resources will require heavy investment also.

A considerable increase in nuclear production, which amounted to only 13 mtce in 1974 (but has nearly doubled since), will certainly be achieved by 2000, though it may well fall short of the 800 mtce target just mentioned. The nuclear output now expected in 1985 – about 150 mtce – is much less than was originally planned. With 1985 only seven years away, construction of new nuclear capacity should have been started, or at least have reached final planning stages by now, if it was to make any contribution in that year, and delays in construction could mean that even 150 mtce will not be attained.

Let us assume, however, that the nuclear and other production forecasts are attained. This will still leave the rest of the EEC with an overall energy deficiency which will have to be met by imports of around 1000 mtce by 1985 and of some 1250 mtce by 2000, the greater proportion of which will be oil.

Britain, as a member of EEC, may be making net exports of energy to her Continental partners in 1985 but by 2000 our North Sea reserves of oil and gas will probably be past their peak. All this makes it essential for the Community to take urgent steps to develop all available energy resources, including coal in particular because of its reserves.

In 1974, the rest of Western Europe – the Scandinavian countries, Austria, Switzerland, Spain, Portugal and others – was proportionately nearly as dependent on energy imports as Continental EEC.

Hydro-electricity plays a more important part in this group of countries than elsewhere in the world and these resources will be further developed, though they will probably account for a gradually reducing share of total demand. Norway will develop her North Sea oil and gas but Norwegian policy seems clearly to be to avoid over-rapid exploitation of these resources.

Several of these countries have moderate resources of coal and their production is expected to increase, especially in Spain. Some may be expected to build up their nuclear power to occupy an important place in their energy balance sheets.

Yet, even on an optimistic assessment, these countries are likely to have to depend on net imports for about a third of their future energy requirements. Net imports were about 230 mtce in 1974 and although this may fall below 200 mtce by 1985, it could rise again to around 300 mtce by 2000.

Taking Western Europe as a whole, including the EEC (with the UK) and the rest of the area as defined above, the period at the end of the century is likely to be fraught with difficulties and uncertainties. If a serious attempt is to be made to insure against these risks, decisions must be taken now.

United States Role

I have emphasised in preceding chapters the important role of the United States in the world energy problem and this is particularly true when we consider the exploitation of its resources. Though the United States is one of the world's biggest producers of oil and by far the biggest producer of natural gas, these reserves have been heavily worked for many years so that future exploitation is bound to pose problems, making it certain that there will be a continuing, and perhaps growing, requirement for imported energy and, because of the scale of US consumption, the claim on the world stock will have serious repercussions for other countries.

Nevertheless, the Americans are a resourceful people and, when determined on a course of action, can motivate themselves and mobilise their skills to a remarkable degree. One cannot but wonder what would be the effect on United States and world energy prospects if a programme to exploit their indigenous resources were mounted on the scale of the space programme of recent years!

President Carter's policies may provide the motivation necessary to bring about a dramatic change in the American energy scene. The most recent available detailed official review of prospects in the United States is the Federal Energy Administration's *National Energy Outlook, February 1976*. This was an attempt to analyse future trends within a consistent framework and to show the likely effects of policy actions which could be taken.

An OECD study *World Energy Outlook* made extensive use of the American publication in presenting its own estimates of US demand and supply, although, in certain areas, the OECD judgment differed from that of the Federal Agency Administration – taking a somewhat more pessimistic view of US prospects.

Both these studies are primarily concerned with the period up to 1985 but it is our purpose in this book to look further ahead to 2000. Estimates for this year were provided in *United States Energy Through the Year 2000* (*Revised*), by the US Bureau of Mines in December 1975. Although both these American publications are becoming out of date and although there are differences between each set of estimates, it is possible to obtain some impression of the way it is thought the US energy scene might develop from the three publications mentioned, taken in conjunction with the forecasts behind President Carter's energy plan – which again, however, concentrate on the year 1985.

Among the conventional fuels, coal is likely to show the greatest growth in the US between now and the end of the century. By 1985, coal production might be nearly 50 per cent above the 1976 level of about 600 million tonnes and well over twice 1976 output by 2000. To attain these levels, however, it will be necessary for the producers to be satisfied, soon enough, that a big expansion will be consistent with other legislation, in particular environmental restrictions on strip mining and on the emission of sulphur from coal burning installations. Within the total I have given, solid fuel exports could expand; but much will depend on internal demand and policies within the United States.

Turning to oil, although US production has been falling since 1970, the balance of opinion is that the rate of discovery of new reserves could be stepped up and increased production achieved if the necessary financial conditions and leasing arrangements obtain from now on. The resultant increase in production may not be large – probably about 20 per cent above the 1976 level of about 700 mtce by 1985 with perhaps a small further increase by 2000.

There were earlier expectations that shale oil could play a significant part by 1985 but, despite the considerable reserves "in the ground", as we saw in earlier chapters, it now seems improbable that oil from this source will play a significant part by 1985, and, even by 2000, any contribution is likely to be small, because of continuing technical and environmental problems.

US oil imports jumped from 440 mtce in 1974 to about 530 mtce in 1976. With increasing demand, present prospects are that they may need to reach around 750 mtce by 1985, rising to perhaps 1150 mtce by 2000.

Natural gas, as we have seen, has been on a declining production trend since the peak reached in 1973, although in 1976, US natural gas production was higher than that of either oil or coal (using United Nations conversion factors). Here, as with oil, improved financial conditions and leasing arrangements for new exploration could probably lead to a rise in production after a further decline, but, even so, production in 1985 may be a little lower than now. Production in 2000 could be rather lower still, in the light of the latest views of the US Geological Survey on total resources of natural gas.

United States imports of natural gas – 33 mtce in 1974 and the same in 1976 – may have doubled by 1985, much of which will probably come by pipeline from Canada, but, by the end of the century, imports could reach 200 mtce, of which the bulk will need to be imported in the form of liquefied natural gas (LNG) from the oil exporting countries. By that time, many other countries will be looking to the same source for their import requirements.

Nuclear power prospects in the United States will depend on the way President Carter's policies in this field are applied. In 1974, the United States accounted for half the non-Communist world's production of nuclear power with 42 mtce* and her share of the total was growing rapidly. U.S. nuclear power was two-thirds higher by 1976 and will certainly expand very considerably but both the Federal Energy Administration and the OECD have indicated that the 1985 total will be well below target and similarly below the forecasts in *Energy Through the Year 2000* (*Revised*).

Our tentative estimates of nuclear power output were 300 mtce in 1985 and 1,000 mtce in 2000. These estimates reflected recent prospects but slippages during the last year have made the 1985 figure look optimistic and technical and environmental difficulties could hold back progress by 2000 to a level far below the figure given.

The USA also produces more hydro electricity than any other country in the world, but it only accounts for about 4 per cent* of the country's total use of energy and, although hydro capacity will expand, it is unlikely to grow as a proportion of the whole.

I should emphasise that the tentative figures I have given for future US energy production, like most other forecasts in this and other chapters, are based on prospects as they have appeared recently. If President Carter can set and maintain in motion a series of effective measures in line with the objectives of his Plan, US import requirements might be less and the remainder of the energy-importing world would be that much better placed to obtain its requirements at more reasonable prices.

Inter-Dependent World

What happens on the United States energy front from now on is of vital importance to us all in an increasingly inter-dependent world in which each country must also consider the other big energy producers and consumers. Canada, for example, is a sizeable producer of energy in all its forms, ranging second to the United States among OECD countries in the production of oil and hydro-electricity, and third, close behind Holland, for natural gas. In 1974, in spite of its own heavy requirements, Canada was a small net exporter of energy, mostly to the United States, with whom it has a considerable two-way traffic.

Unfortunately, exploration for further oil reserves in Canada has had relatively disappointing results in the recent past and production from

*It should be explained that this figure is arrived at by using a conversion factor which reflects the relative importance of nuclear and hydro power and not that used by the United Nations Statistical Office.

its vast reserves of tar sands, details of which were given in an earlier chapter, continues to be beset with difficulties.

Canadian oil production has fallen by a quarter from its 1973 peak and may fall a little further still by 1985, leaving Canada a net importer of oil to the extent of, say, 80 mtce.

Production of other fuels will probably increase, though not enough to match the growth in overall Canadian energy requirements. Canada may continue to be a net exporter of natural gas because there are demand centres in the United States which are nearer to Canada's sources of supply than are her own Eastern States.

Looking beyond 1985, much will depend on the discovery rate of new oil reserves. Unless this improves, Canada could be importing 200 mtce or more of oil by 2000, which would be a big addition to the demands likely to be made on OPEC countries by other countries by that date.

Japan has little in the way of indigenous resources, compared with the needs of its bustling economy. Its coal output has shrunk to under 20 million tonnes a year and is now less important than hydro electric power, at present at 30 mtce (on the conversion basis referred to in the footnote on page 96). Hydro power will probably expand further but cannot ever be more than a marginal item in the energy balance sheet. The future of Japanese coal is uncertain but it may continue at about its present level. Geothermal power may make a contribution because Japan is a volcanic country in origin, but it is not likely to reach the scale even of Japanese coal. Solar, wave and wind power may all have a contribution to make later though probably not much before the end of the century.

Hopes of keeping energy imports within reasonable bounds thus rest heavily on nuclear power for which Japan has ambitious plans, but which face, there as elsewhere, serious environmental opposition. It will be difficult enough even to find the necessary sites in the crowded Japanese islands without the added difficulty of the strong opposition which shows no signs of diminishing.

Japan's plans for nuclear power have been drastically cut back – the latest estimates show another big cut for 1985 – but all the same we can assume that there will be a big expansion over the years beyond stations already in operation or under construction. On present prospects nuclear power might reach 40-50 mtce in 1985 and perhaps 250 mtce by 2000. This would mean that for the remainder of the century Japan would make a continuing major demand on the world's reserves of exportable energy.

As an indication of the scale of the problem, Japanese energy imports are running at around 450 mtce a year of which oil represents about 380 mtce. The total could approach 800 mtce by 1985 and half as much again by the end of the century of which, by then, over 800 mtce could be in the form of oil, with a considerable import of natural gas. Both these fuels will only be obtained from the sources on which the rest of the importing world will depend.

Although Japan's oil imports overshadow its intake of other forms of energy, it is, nevertheless, by far the world's biggest coal importer. Up to now imports have been almost entirely of coking coal for the steel industry but future imports are expected to include steam coal as well. Coal imports could rise from the 1976 total of 61 million tons to half as high again by 1985 and perhaps double by 2000.

Other countries which may be considered in relation to the world's energy balance include for example Australia and South Africa which are already exporters of sizeable quantities of coal and could increase their exports further. Yet others, such as Mexico, have been importers (of oil) but are likely to become exporters because of new developments. India, because of its size and in spite of a low *per capita* consumption, uses a lot of energy but, at the moment, produces most of it indigenously. If its energy usage increases greatly it could well be based on an expansion of indigenous supplies.

Though many of these and other countries outside the major areas we have been mainly considering are at least partly self-sufficient in energy and their indigenous production is expected to increase substantially, taken as a whole they imported about 270 mtce of oil in 1974 and might need imports of over 400 mtce by 2000.

Available for Export

Whatever the future may hold on the energy front, there will be a continuing demand from Western Europe, the United States, Japan and other countries for imported energy, in the form of oil, natural gas or coal according to circumstances.

To take oil first, the tentative forecasts used here imply that the importing areas will want to import about 2700 mtce in 1985, against about 2100mtce in 1974, and that their import requirements might reach 3800 mtce by 2000. (See Table 8) These figures, of course, are based on the assumptions given in Chapter 4—a doubling of real prices, a reasonable rate of increased economic growth and of energy conservation.

There is little doubt that the OPEC countries, with assistance from other exporters, could provide 2,700 mtce, in 1985 and they could probably meet the increasing requirements up to the early 1990's. Yet there must be very great doubts as to whether they will be able to expand to meet a demand of 3,850 mtce of oil by 2000. Certainly there is little likelihood on available evidence that they will be able to meet a continuing increase in demand after 2000.

We have to ask ourselves also whether the exporting countries will wish to meet such high requirements, even when it would be possible for them to do so – and at what price they will be prepared to sell their oil. OPEC's monopoly position, and the pressure of demand for oil, will give exporters considerable freedom to raise prices and to decide what supplies to make available, always bearing in mind, however, that it cannot be in their interests to provoke major economic crises among their customers.

Table 8. Tentative Forecasts of Oil Imports Required (*mtce*)

	1974	1985	2000
Western Europe	1029	1000	1200
United States	440	750	1150
Japan	378	650	850
Other areas	270	250	650
Total	2117	2650	3850

Generally, it is likely to be those countries with limited oil reserves which will be keenest to restrict production and conserve resources. Some OPEC members, such as Iran (the second biggest producer) urgently need more foreign exchange for major economic and industrial development, and also need rapidly increasing amounts of oil for their own internal use.

If the gap between oil import availability and demand is not closed by other means it will be closed by the operation of the price mechanism, which will mean that by 2000 prices could be much higher than the doubling in real terms we have assumed. This could only in my opinion, as expressed elsewhere in this book, be averted by the importing countries developing vigorously the fullest economic range of indigenous resources they have available and stimulating strong conservation measures.

A factor which could affect the future development of world resources would be the entry into the world oil export market of countries which are not at present selling overseas to a significant extent. The USSR is the obvious example, with sufficient oil in the ground to increase its exports considerably though, as I have mentioned when discussing world reserves, much of the oil will be difficult and costly to extract and, in any case, may be required to meet rising internal demand and that of other Communist countries.

I mentioned the recent upgrading of Mexico's reserves in Chapter 3 which, if the estimates of probable reserves prove to be accurate, could add to the world's exportable oil resources. It is too early to give precise figures of likely availabilities but they are unlikely to make a major difference to world supplies.

Another unknown factor in the oil import-export equation is the possibility of increased exports from China, which is already shipping small quantities to Japan, the obvious market if the People's Republic is looking for more foreign exchange. On the other hand, China's internal demand for oil is bound to rise with increasing industrialisation so that future exports must remain very much a matter for conjecture.

I therefore conclude that, although the USSR and China and, on a smaller scale, some other countries not previously mentioned but which could fortuitously come up with unexpected discoveries, may postpone the day when the import-export gap appears, I see no reason to modify the widely held view that effective action to reduce oil import requirements must be put in hand now.

Natural Gas

Exploitation of the world's natural gas reserves is partly affected by similar conditions to those which apply to oil (since many of the biggest reserves are also in the Middle East and Africa) and partly by the technical and economic problems of expanding the trade in natural gas from the relatively low present level to the considerable quantities which seem to be envisaged in the consumption forecasts of the major consumers.

Natural gas exports from OPEC countries amounted to around 20 mtce in 1974, at which time the USSR was actually a net importer. Holland supplies gas to the other members of the Community while Canadian sales to the United States are the biggest single movement of natural gas across the frontiers anywhere in the world at the moment.

As we saw in an earlier chapter, there are considerable reserves of natural gas around the world but some big questions hang over the future exploitation of these resources.

Locally, within a particular region where reserves exist, there will be scope for increased trade in natural gas. North Sea gas, for example, will be fully exploited. Gas from the British sector has so far not been exported but Norway is selling gas from her Frigg field to Britain and reserves in the Ekofisk field, also in Norwegian waters, are under contract to a consortium of German, Dutch, Belgium and French gas authorities.

These supplies will contribute to Western Europe's big anticipated demand for natural gas. The remainder of the region's expected requirements for 1985 are mostly already under contract in the form of supplies from Algeria, the USSR, Iran and Libya. Whether these sources, and others with gas available, will be able and willing to meet the doubled requirements anticipated for the year 2000 is more problematic. The necessary gas, as we have seen, exists in the ground, but Western Europe will, as with oil, be competing for the available supplies with other countries.

Prominent among the competitors will be the United States and Japan. To meet all the demands will require the extensive exploitation of Middle East and African reserves. The Soviet Union figures in US plans, too, with suggestions of American participation in developing the USSR Siberian reserves. If this should develop, it could adversely affect Western Europe's changes of increased Soviet supplies. Contracts are in hand for the export of L.N.G. in considerable quantities from North Africa to the US.

Japan has also been busy arranging contracts for L.N.G. imports and aims to increase her "take" considerably in the years ahead. It already imports from Brunei and will be getting more from South East Asia but a lot of its prospective requirements are likely to come from the Middle East, to which the rest of the world will be looking.

Unless the USSR exports a lot more gas than now seems likely, demands for gas from the Middle East and Africa might exceed 500 mtce by 2000, on our tentative assessment of import requirements. I have some doubts as to whether a total of this order will be achieved in practice although these large reserves have not yet been exploited to any great extent.

Exports from these areas will obviously rise substantially during the period but the extent to which the reserves are developed will depend on the willingness of countries to do so. Those OPEC countries with a lot of gas to sell compared with oil will tend to price their gas just high enough to sell at comparable world prices while others, with more oil than gas, will have no incentive to undercut their own oil and will plan the exploitation of both oil and gas as a related exercise. The extent to which the USSR expands its gas exports is likely to depend on how much it needs to increase its foreign exchange earnings.

There is a great deal of conjecture here but whatever the outcome, it will demand a great deal of physical effort, to say nothing of the financial implications, to produce the necessary pipelines, refrigerated L.N.G. tankers, liquefaction and re-gasification plants which will be required. Thus, while I expect trade in natural gas to expand considerably during the remainder of the century, supply may fall short of the demand levels indicated above, even at the high energy prices which will be ruling by then.

Coal Again

This brings me once again to the role of coal which, as the world's most abundant fuel, should be in a position to make a significant contribution to world trade in energy, provided the coal industries exploit their reserves to best advantage. With present-day large capacity ocean-going bulk carriers and highly mechanised loading and unloading techniques it would seem to be a relatively simple matter to transport some of the abundant coal from places where there are big reserves to the markets, in Western Europe and Japan in particular, where energy is going to be increasingly scarce. Unfortunately, reality is hardly as simple as that.

Massive investment in production and transport facilities is required and the time span for a big increase in supply is long. In those countries where there has been a rundown in coal production over many years in the past, with a corresponding lack of investment, a great effort will be needed even to maintain, let alone increase output in an extractive industry like coal. Many countries, even if they have big reserves, will be hard put to it to meet domestic demand. Two countries which have the coal in the ground to step up coal production sufficiently to meet their own demand and also provide great additional quantities for export are the United States and the USSR.

As we have seen, the USA is likely to increase its coal output substantially provided it is able to overcome environmental and other problems, but the great proportion of the extra output is likely to be used for her own consumption. An increase of US exports by 2000 to, say, double the current level of around 50 million tonnes is possible – enabling the US to retain its position as the world's biggest exporter of solid fuel – but anything substantially higher is unlikely.

Much of future demand for coal, both in the Community and Japan is for coking quality, for which the US has hitherto been the main international supplier, with exports around 40 million tons a year in recent times. It seems very likely, however, that more and more US coking coal production will be diverted to home use, including power stations anxious to use this type of coal to comply with sulphur content regulations, as well as the increasing demand from the steel industry. Exports of coking coal to Europe may therefore decline, while price levels rise.

The USSR's net exports were 20 million tonnes in 1974 and with vast coal reserves it might be able to expand exports in the future. The USSR's problems are of a different nature to those of the USA, since they are related to the nature of the locations where the coal lies rather than to environmental objections. More than 90 per cent of reserves are in the Asiatic sector, predominantly in Siberia, and coal from these areas would have to travel great distances overland to reach Western Europe which, in 1974, received 8 million tonnes from the USSR. Nor would Japan be any easier to supply.

Rapid expansion of Soviet output is planned but it will need a great effort to solve the transport and other problems of remoteness and climate, and the bulk of any additional output is likely to be required for internal use. Some increase in exports is probable because of the attractions of the extra foreign exchange, but any such exports are unlikely to be large enough to have any great influence on the world energy picture.

At present, both Poland and Australia export more coal than the USSR and this seems likely to continue. In 1974, Polish solid fuel exports were 42.5 million tonnes, half of which went to Western Europe while Australia exported 29.5 million tonnes, including large sales, (23 million tonnes in that year) to Japan.

Both these countries export coking coal and Australia will undoubtedly continue to exploit the Queensland reserves where new pit-to-port facilities are under development.

Australian coking coal is likely to be available in increasing quantities for Western Europe which, in 1974, took 5 million tonnes of all forms of solid fuel. This will certainly rise in the years ahead due to the high quality of Australian coking coal. Judging by a paper presented to the 1977 World Energy Conference, Australia's coal exports could well increase by more than those of any other single country.

It has been reported that the Poles aim to double their total coal production by 2000, compared with 1972, when it was 150 million tonnes. Total exports could then be 70 million tonnes, or even more. In periods of low demand the Poles tend to price coal below estimated costs, in order to earn foreign currency. In periods of strong demand they react correspondingly to the market situation.

Other countries which are likely to make a continuing contribution to world trade in coal include South Africa and Canada. South Africa will want most of her output for herself but her 1974 exports of 1.5 million tons could potentially increase tenfold by 1985. The position in 2000 could be more problematic. Canadian coking coal deposits are located some 400 miles from the Pacific coast, so that heavy internal transport charges must be added to the sea freights from West Coast ports.

A large part of Canadian exportable production is committed under long term contracts to the Japanese steel industry. Moreover, Canada also imports coal from the USA into her Eastern states and is only likely to be a minor net exporter – in fact, in 1974, she was actually a net importer.

Other countries which have been mentioned as possible sources of future coal exports to Japan include China, Indonesia, India and Mozambique. None of these are exporters of any size at present and even if they do start supplying Japan, the amounts are likely to be small for some time.

After looking at all sources of coal exports we must conclude that, on present prospects, although they are likely to increase, they will only make a relatively minor contribution towards the world's total energy requirements. Nevertheless, substantially more coal is likely to be used in the producing countries themselves thus reducing the general pressure on internal supplies.

Exploiting Britain's Reserves

I have left it until now to consider the question of the exploitation of Britain's energy resources, partly because I felt it advisable to present the world scene first, as the background to Britain's role, and partly because our position, for the time being at any rate, is more fortunate than many of the other members of the non-Communist world. The fact that Britain possesses abundant coal to exploit for as far ahead as we can see, while we can also exploit our indigenous supplies of oil and gas, puts us in this exceptional position.

I have already set out the latest position on North Sea oil reserves in Chapter 6. Five oilfields came on stream in 1976 to add to the two which started up in 1975 and production has already reached a rate equivalent to more than 40 million tons a year. The UK should reach self-sufficiency in oil by 1980 or thereabouts, with output in the early 1980's being within the range 100-150 million tons a year.

What happens beyond that is less certain depending partly on whether resources come up to expectation and partly on what actions Government takes. Each individual field goes through a cycle which starts with increasing production for two or three years or so up to a peak which may only last for a similar period before a decline sets in. Eventually a point is reached when further production is either economically unproductive (depending on prevailing oil prices) or technically impracticable.

At present most fields are in the first part of the cycle so the total output rate is rising fast. The momentum of development of fields already discovered, including those yet to start producing, is likely to mean that output will exceed our internal requirements in the early and mid 1980s, leaving a surplus for export. But once the older fields have passed their peak, output can only be sustained or increased if sufficient new fields come forward. Eventually the rate of discovery of new fields will fall below the rate needed to hold output steady and the total output rate will decline. To some extent the time when this will happen and the rate of decline can be controlled by licensing policy and perhaps by restricting development of fields in areas already licensed. The higher the overall peak output is allowed to rise in the early years of exploitation of the North Sea the sooner will the decline set in and the faster will be the rate of fall.

The Government will therefore have a choice between continuing to exploit the North Sea rapidly, and so achieving a considerable level of oil exports in the mid 1980s or restraining expansion so as to keep more for later years when the situation can be expected to be more difficult. The Government has said it will not impose production limits before 1982, but has taken no decision beyond then, and has not announced its policy on future licenses. Exports beyond the minimum necessary for operational and balance of payments purposes would mean increased revenue earlier at the cost of a loss of security later.

Whatever the policy adopted for oil exports it is likely, on present knowledge of resources, that Britain's oil production will be declining by the end of the century. If production has been virtually unrestricted during the 1980s and early 1990s it will probably have fallen far below our own internal requirements by 2000, and substantial imports will be wanted–obtainable only at heavy cost. If, however, production has been kept down close to internal requirements Britain might still be self-sufficient in oil up to about 2000.

North Sea gas faces much the same situation. The earlier fields in the Southern North Sea are fully developed and are either at or about to reach the plateau from which the inevitable decline must start in a few years. Gas from the UK/Norwegian Frigg field will increase the supplies available to the British Gas Corporation and there is also likely to be a lot of gas produced in association with oil coming from Brent and other oil fields in the Northern Basin of the North Sea – especially if the gas gathering pipeline to which reference has already been made, proves to be a feasible proposition. As with oil, the Government is likely to have some choice on how fast to use up the resources. It would be wasteful to use this premium fuel for crude heating purposes where coal could do the job well.

Natural gas production, could possibly be on the decline by the end of the century. When the decline starts and how fast it is will depend to some extent on how far the Government has controlled its expansion in the meantime – and also of course (as with oil) on the extent of future discoveries.

Even if the Government does exercise a considerable degree of control over oil and natural gas output during the next 20 years, and so conserve resources for later use, it is clear (on any assessment of North Sea resources that looks possible now) that we shall need greatly increased amounts of coal and nuclear power by then if there is to be any chance of avoiding heavy imports of oil. The need will be all the greater if only a limited degree of control has been exercised or if future discoveries are disappointingly low.

Regeneration of Coal

North Sea oil and gas are both industries which, thanks to a massive injection of finance and considerable technical skill, have been brought quickly to a high pitch of development, providing a solid base from which to exploit the reserves known to exist, and still to be found, beneath the surface.

Coal, on the other hand, is an old-established industry which, having been allowed to fall into decline, has needed a period of regeneration in order to give it a platform from which to exploit energy reserves many times those of offshore oil and gas combined. Fortunately, we have the basis of our regeneration in our forward planning, which is already being translated into action. The problem now is to maintain the momentum we have gained and steadily build up the scale on which we exploit our coal reserves during the rest of the century, and beyond.

'Plan for Coal', launched in 1974 and the more recent but complementary 'Plan 2000', outlined in Chapter 3, represent the blueprint for the future exploitation of reserves. In the shorter and medium term it involves raising output by extending the life of existing mines and new sinkings on present coal fields. Longer term it will be necessary to open up completely new coal fields.

Because of the importance of this latter side of our activities to the future of the coal industry and therefore to the country's energy prospects I propose to deal at some length with the way in which we set about discovering and then exploiting a new coal field.

Exploration methods include the drilling of boreholes which provide cores of the coal measures, and the use of modern seismic techniques.

Having selected an area for exploration, on the basis of broad geological evidence, the first stage is to sink one or two boreholes, sufficient to confirm the existence of any mineable coal seams which may be present. Basic information can be obtained on the thickness of the seams and the spacing between them, the type of coal, its ash content and other relevant information, but this data relates only to the very limited number of boreholes sunk.

A second exploratory stage therefore follows, with the drilling of more boreholes, spaced out to determine the extent of the discovery and the likely quantity of mineable coal in order to decide whether the amount which can be extracted will justify the expense of a completely new mine.

In those cases where the initial stages of exploration are encouraging, there follows the third stage when a programme of closely-spaced boreholes, ranging from one to three miles apart, depending on the nature and structure of the reserves, is carried out. This can entail the sinking of another 20 boreholes to confirm each additional colliery site capable of producing 2 million tons of coal per annum. At the same time, some 75 miles of seismic survey may be required, covering a grid pattern over the reserve area. This information, plus the boreholes, gives a three-dimensional picture of the area, enabling the future mine to be planned in the confidence that we shall avoid the geological disturbances below the surface which have often caused problems in the past.

Finally, still more boreholes are drilled at specific points in order to provide detailed engineering information on which to position mine shafts and drive the roadways, or drifts, along which the coal will be extracted.

Experience in the use of boreholes and seismic techniques has been growing all the time and the discovery and delineation of the Selby coalfield in Yorkshire is the classic demonstration to date of the success of these methods. Here, too, we followed the policy of only initiating the exploration programme in an area where it was geologically fairly certain that coal existed.

As long ago as the early 1900's a few boreholes drilled in the area to the south and south-west of Selby had shown no seams of economic thickness, but drilling in the area after nationalisation had suggested that the previous borehole findings could well have been erroneous. Selby, then, was no "wildcat" adventure.

Nevertheless, the finding of the huge reserves we have now proved – at least 300 million tons – reads rather like a detective story. Between 1964 and 1967, five deep boreholes drilled west and north of Selby had proved conclusively that the boundary of the concealed coalfield which was thought to exist turned from the east-west direction in which it had seemed logical to pursue the original search, to a north-south direction, thus giving rise to a north-east extension of the Yorkshire coalfield.

These boreholes also added another twist to the detective story. It concerns the Barnsley seam (coal seams are often named after the area where they are first located), traditionally the best of all Yorkshire coal. This seam had previously been proved to split into several layers (or thin seams) south of Selby but the boreholes now proved that the layers in fact came together again further on to form a thick seam of this high quality coal at workable depth.

Although I may have seemed to have made it all sound deceptively simple, the actual procedures adopted at Selby involved technological advances which will be cumulatively beneficial in future exploratory work. This is particularly true of the seismic techniques employed. These have been used in the coal industry for many years, as a modified form of those used by the oil industry. In this form they had been employed since 1959 with only moderate success in detecting structural faults.

However, in 1973, a system was developed at a borehole site in the centre of the Selby coalfield which yielded records some five times as good as the best results previously gained, and at only about twice the cost per mile. This additional cost is more than justified by the quality of the information obtained.

Exploration techniques developed and applied at Selby, are now being used elsewhere in our programme. Although Selby with its eventual 10 million tons a year capacity (the largest of its type in the world we believe) has understandably caught the public imagination, the Selby mining complex will only account for about half the 'Plan for Coal' production from new mines.

For the other 10 million tons, we are looking to such prospects as Park, Staffordshire, where exploration has proved the existence of some 130 million tons of reserves, capable of supporting an annual output of at least 2 million tons, and North East Leicestershire where a programme of about 50 boreholes has proved the existence of four thick coal seams at a workable depth. Here are some 500 million tons of readily workable reserves.

To prove the existence of reserves is only a part, though an important part, of the exploitation of our coal resources. There remains the extraction of the coal by the most efficient methods, at economic cost and – a particularly vital question which I shall deal with in more detail in Chapter 9 – without causing unacceptable disturbance to the environment.

At Selby, we believe we have developed such a method of extracting the coal. Briefly, it involves the sinking of five shaft mines in different parts of the field, each to produce 2 million tons of coal a year, though the coal will not be brought to the surface at these sites but will be transported on huge underground belt conveyors, occupying a central spine system, surfacing by a drift at existing railway sidings at Gascoigne Wood. Here, the coal will be loaded into "Merry-Go-Round" (MGR) trains, each capable of carrying 1000 tons of coal at a time to power stations at Eggborough, Drax and Ferrybridge, all within a 10-mile radius.

This necessarily brief outline of the Selby project does scant justice to the many technical and engineering innovations which will be incorporated in what we believe will be the first of a new generation of coalmines, not only in Britain but in the world.

For example, there is the scale on which everything is being conceived, from the five shafts, which will accommodate cages capable of transporting 90 men at a time and major items of machinery, to the high capacity conveyors which will bring up between 40,000 and 50,000 tons of coal a day. Every aspect, both of underground operations and surface coal handling, will be regulated by a central computer which, through a sophisticated data processing system, will control all mechanical functions and also monitor environmental conditions throughout the mine complex. Equally up-to-date will be the comprehensive communications system, including the use of a new process of underground short-wave radio, which will ensure overall supervision of the whole vast project.

Plan Into Action

To translate a scheme of the magnitude of 'Plan for Coal', of which the Selby achievement is only a part, demands heavy investment, on a level far higher than that to which the industry was accustomed before 1974 when the amount of investment on new major prospects had been as little as £7 million a year. The Selby project alone will cost over £400 million (at 1976 prices) spread in varying proportions over ten years.

As an indication of the increased effort expended since the first three years after the issue of the Plan in 1974, no fewer than 80 projects have been initiated at existing mines, representing additional output of 16 million tons by 1985 – or some two-thirds of the Plan total of 22 million tons from existing mines.

Equally good progress has been made on the section of the Plan which deals with completely new mines. By the end of 1976, over 11 million tons of additional output was either in operation, under construction or approved – over half the 1985 total of 20 million tons from new mines envisaged in the Plan.

A good start has also been achieved in the increase of open cast production, suggesting that the increased production of 5 million tons indicated in the Plan will be reached by the early 1980's.

Yet, although the rate of advance on all fronts has been impressive, anything less would have been regarded as unsatisfactory because of the long investment lead times – over 10 years in some cases. Because of this great length of time needed for exploration, planning procedures, and investment cycles, decisions must be taken for new mining capacity still further ahead than the 1985 goals of 'Plan for Coal'.

Hence 'Plan 2000', which continues our strategy from 1985 up to 2000, has been devised, to prevent diminishing capacity later in the century. Our thinking is based on the assumption that for the greater part of the rest of the century the major markets for coal will continue to be found in power stations, coke ovens mainly for the steel and foundry industry, in many general industrial sectors and as a domestic fuel. Towards the end of the period however, with the reduced availability of our indigenous oil and natural gas, coal is likely to be required in increasing quantity for conversion in order to supplement supplies of these fuels. Hence the major research effort which will be described in the next chapter.

Summing up the UK situation, we estimate that the demand for coal by 2000 will fall somewhere between the 135 million tons on which 'Plan for Coal' is based and 200 million tons a year. A reasonable assumption would be an annual total around 170 million tons, that is, in the middle of the range, made up of 150 million tons of deep-mined coal and 20 million tons from open cast sites.

What are the implications of this for future exploitation of our coal reserves? It offers a formidable challenge since, on the assumption that existing capacity exhausts at about two million tons a year, it will require some 100 million tons of new capacity to achieve the proposed 150 million tons of deep mined coal by 2000. Taking the 25 years from 1975 to 2000 this will entail a rate of opening up of new capacity averaging 4 million tons a year over the whole period.

After 1985, most of this new capacity will have to be in the form of new mines, since we expect to have utilised the major possibilities of more capacity from existing mines by then. Put another way, it will mean that two-thirds of the industry's capacity will have been renewed in 25 years – one of the most striking examples of industrial renewal in recent British economic history.

Exploiting Other Resources

Britain's prospects for exploiting her resources of fossil fuels for the remainder of the century can be defined with reasonable certainty, as we have seen. It is much less easy to plot the path ahead for the other forms of energy, particularly nuclear power.

As I write this chapter, the question of Britain's future nuclear policy is still undecided, beyond the fact that the present programme of Advanced Gas Cooled Reactors (AGR's) should bring our nuclear production up to about 25 mtce by the early 1980's.

The whole question of future reactor choice and the scale of future nuclear development has still to be decided. The uncertainty is partly due to technological problems and the costs and relative advantages of the different types of reactor, not excepting the questions of safety which loom particularly large in the public mind.

Doubts have also arisen about the number of new power stations, nuclear or otherwise, which will be required, in view of the fact that electricity demand is not now expanding as rapidly as in the fifties and sixties (when it was doubling every ten years). In the last few years demand has been depressed by the recession and by the extra natural gas which households have been taking. Growth in demand will pick up again when the economy improves but it is unlikely to average anything like its former rate. Whatever the level of demand however it is going to be essential to have a sufficient proportion of both coal-fired and nuclear-powered stations – and this will mean new orders for both types. In the case of coal, as distinct from nuclear, present capacity is relatively high, but much of it is old. If the old plant were retired without replacement by other coal plant after 30 years of life, by the end of the century the amount of coal-fired capacity would be down to a quarter of its present level. For this reason we have set great store on having orders placed for new coal-fired plant and welcome the decision to proceed with the completion of the Drax station.

To bring to an end our consideration of the exploitation of Britain's energy resources, I must make a brief reference to the "benign and renewable" forces of wind, wave and solar power. Reports issued by the Department of Energy have shown that it would be technically possible for the renewable sources to make a contribution of perhaps 25 mtce by the end of the century. This would represent some 5 per cent of total expected demand at that time. Much effort would however be required to achieve this.

Wave power is perhaps one of the most promising, though it is still far removed from the stage of commercial application. Another form of water power, the Severn Barrage, remains under consideration, well over half a century since it was first mooted. It could, it is reckoned, provide up to 10 mtce a year but the cost would be heavy, the construction time long and there would, once again, be environmental problems to overcome.

Solar energy is probably as far advanced as any of the renewable sources and can help to cut costs for low temperature hot water in houses. A good deal more development effort is however required before it can be widely used. Wind power and geothermal power are likely to have more limited scope in Britain.

Britain's exploitation of its indigenous resources will thus rest heavily on the use made of the relatively limited oil and gas reserves while developing the abundant coal reserves, both to eke out the oil and gas and to make a major contribution to overall energy demand when the flow of these two fuels starts to decline.

8 The Vital Role of Research

In the previous chapter I showed that the world's reserves of energy, at present locked up in the earth, beneath the sea and, indeed, in such natural forces as the wind, the waves and the sun, will fail to meet our growing needs for the remainder of the century, and beyond, unless they are fully and correctly exploited. This does not just involve putting greater effort into existing methods of extracting and using our energy sources but also demands a significant stepping up of research and development programmes on all aspects of energy.

I say "stepping up" because research and development has always been a continuing activity of the world's major fuel industries and I have no intention of belittling the work which has been done over many years and which has resulted in the steady improvement in the quantity and quality of the fuels from which we have benefitted in the past.

What may have been adequate in the past will no longer serve in the changed situation which has emerged and which, as I hope I have made clear, demands a totally new approach. So far as research and development are concerned, governments and peoples must accustom themselves to devoting a bigger share of available resources to these activities than in the past.

Nor is this something which can be put off till tomorrow, because, as with so many other aspects of the energy problem, the time scale is long. Just as we are looking ahead to 2000 AD as the target date for the exploitation of reserves, so must we also start to plan now for research projects to come to fruition by the end of the century, though if early results begin to show between now and then, this will be a welcome bonus. The development of any new energy technology of future significance can be expected to take at least a decade and possibly much longer.

I have emphasised in the earlier part of this book that the energy problem is very much an international matter in which Britain's coal industry and the other energy industries have their parts to play in collaboration with other comparable industries. I adopt a similar approach to the question of research and development, which must have an increasingly international flavour from now on.

Research Begins at Home

For the moment, however, I am leaving international aspects of research on one side because, although it will play an increasingly important role in future, all energy research, like charity, begins at home, in sound, nationally-based programmes related to local requirements and resources.

Whilst there is a need for work in virtually all energy fields, there is a particular urgency in research into the extraction and utilisation of coal, which as the world's largest source of fossil fuel, will have to bear the brunt of the battle for energy for the foreseeable future.

In the United States, for example, which has by far the world's biggest energy research and development programme, coal is already receiving a reasonable share and, under President Carter's new energy initiative, seems certain to receive more. Back in June 1973, when President Nixon announced a major research effort which was to cost 10,000 million dollars over the five years beginning in 1975, increasing it in January 1974 to 11,300 million dollars, the annual average share of coal (580 million dollars) represented a seven-fold increase over the figure for 1973. Coal had the largest share of the budget apart from the proposed annual expenditure of 1,120 million dollars on nuclear research.

Federal energy research in the United States has been in the hands of the Energy Research and Development Administration (ERDA) since it became operational in January 1975, when it took over the energy research and development programmes previously carried out by the Atomic Energy Commission, the Office of Coal Research in the Department of the Interior, this Department's other energy research, the work of the National Science Foundation and that of the Environmental Protection Agency. ERDA's functions are now being acquired by the new Department of Energy.

When he signed the Bill which established ERDA, the President said: "ERDA must and will become a lot more than the sum of its present parts. What is envisaged is nothing less than a complete energy research and development organisation. It will be one which will fill in the gaps in our present research efforts and provide a balanced national research programme. It will give proper emphasis to each energy resource according to its potential and its readiness for practical use. It will closely integrate our energy research and development efforts with overall national energy policy." ERDA's total energy budget for 1977 amounted to 3,000 million dollars of which coal accounted for 400 million dollars and nuclear research 1,650 million dollars.

This was the position before President Carter's initiative on energy, with its objective of an increase of coal production by about two-thirds, accompanied by a major expansion of research, especially into more efficient ways of using coal. Undoubtedly larger sums will now be devoted to this end.

Research expenditure under the ERDA umbrella represents only a part, though much the biggest share, of total United States energy research and development. The fuel industries themselves, being mainly in private hands, also spend considerable sums on research on their own account.

In Western Germany, too, where the energy industries are also mainly privately owned, the Government boosted its research effort in the wake of 1973 crisis, with a new programme covering the years 1974–77. It covers all forms of energy, but coal figures prominently in a programme which includes projects in mining engineering, coal preparation, processing and utilisation.

Federal and provincial (or Land) governments co-operate in the implementation of the programme, under which grants are made for specific projects, the size of the grant depending on the importance of the project in the eyes of the authorities.

Outside Western Europe and the United States, coal research is claiming its share in many national budgets. Australia and Canada, for example, are both expanding their coal industries and, having become major coal producers during the past decade, have supported this progress with a significant growth in research and development. In both these countries research is not sponsored at national level but is the responsibility of individual states, such as New South Wales and Queensland in Australia and British Columbia and Alberta in Canada. In both countries, interested commercial bodies also undertake research into coal.

Research is now recognised as essential in countries which may not have been major energy producers in the past. India, for example, whose vast and growing population will need a big increase in energy availability if living standards are to be maintained, let alone increased, regards coal as an important indigenous energy resource and the aim is to increase production from 88 million tonnes in 1974–75 to 124 million tonnes in 1978–79. Research and development will make an important contribution towards the attainment of this target.

Energy research in the Soviet Union is on a large scale, covering all major fuels, including some notable work on nuclear energy. As far as coal is concerned, the current Five-Year Plan aims to increase the annual output of coal and lignite from 700 million tonnes in 1975 to 800 million tonnes by 1980, a target which will need to be backed by a commensurate research and development effort.

When I visited Poland in 1975, at the invitation of the Polish Ministry of Mining and Power, I learned that a big expansion in research had been in hand for some time and I was told that continuing work on the scientific and technical problems of Poland's coal industry will be accorded a heavy expenditure of resources for the foreseeable future.

This necessarily brief review of research and development in selected countries gives an indication of the work in hand at national level and this must remain the backbone of the attempts of the energy industries to make the best of the available resources. However, as I have already said, I regard the energy problem as very much an international matter. Before I turn to consider what Britain is doing about research, I therefore look at what is happening on the broader world scene.

International co-operation is taking two main forms. On the one hand, there are bilateral research agreements between pairs of countries with common problems, to pool their resources and eliminate wasteful duplication of effort. On the other hand, there is a growing, and welcome, trend towards wider co-operation in research through multi-national arrangements.

Working Together

One result of the 1973 crisis which should prove to be of lasting benefit was the formation of the International Energy Agency (IEA) in the autumn of 1974, as briefly mentioned in Chapter 3.

Included in its broad terms of reference and, indeed, one of its main objectives, is the development of increased energy supplies, from both existing and new sources. To this end, a series of Working Groups was set up to look at research and development aspects of a range of energy sources, including nuclear power, solar energy, geothermal energy and, of course, coal.

IEA possesses no central funds to support any of this work, and its role is purely organisational, bringing together countries with common interests and problems, and encouraging cooperation between them, to further IEA objectives. Particular member nations (as I mentioned in Chapter 3, they are restricted to countries in the OECD) are invited to take the initiative in organising collaboration in the various fields of research and development.

This can be illustrated by reference to coal where, because of its long involvement in coal technology, the British Government was invited to take the lead. The Department of Energy asked the NCB to act on its behalf by setting up the Coal Working Group. This was done in February 1975, with Leslie Grainger (former NCB Board Member for Science) as Chairman.

Although there were no precedents or established procedures for international cooperation of this type, progress was extremely rapid and the Coal Working Group quickly agreed on five priority projects. They included one large experimental plant to investigate the exciting possibilities of fluidised bed combustion (of which more later) and the establishment of four facilities – an Economic Assessment Service, a Technical Information Service, a Data Bank of World Reserves and Resources and an International Mining Technology Clearing House.

Three of these four IEA Services occupy offices in Central London whilst the International Mining Technology Clearing House is located at NCB's Mining Research and Development Establishment at Bretby, near Burton on Trent. The fluidised bed combustion facility, jointly sponsored by the United States, West Germany and the UK, is located at Grimethorpe Colliery, near Barnsley in South Yorkshire.

Each of the four services has an important role in ensuring the optimum use of the world's coal. For example, the Economic Assessment Service will help to identify areas in which new research projects should be undertaken and ensure maximum benefits from the exploitation of new coal reserves and technology. Work will cover many aspects, from the economies of coal utilisation plant and effluent disposal problems to coal transport and its conversion into different end-products.

Developments in coal technology world-wide will be the province of the Technical Information Service which will provide a central source of information for both coal producing and consuming countries.

As its title suggests, the Data Bank of World Reserves and Resources will assemble information for use as a basis for predicting the sequence in which world coal reserves are likely to be extracted and in assisting the development of world trade in coal.

Finally, the International Mining Technology Clearing House will establish a register of research and development work now in hand around the world. This will enable IEA members to discover quickly what comparable work is taking place in other countries, leading to useful technical collaboration or exchange of information in areas of mutual interest.

None of these services operate in watertight compartments but together they provide a comprehensive guide to all major aspects of the international coal industry and will arouse and sustain an awareness of its vital contribution towards meeting future energy needs.

Although the IEA is to be warmly welcomed as an example of recent international cooperation in energy matters, it should not obscure the fact that the coal industries of Europe were pooling their experience, including research and development, long before the 1973 energy crisis focussed public attention on the problem.

As long ago as 1951, for instance, the Treaty which established the European Coal and Steel Community (now part of the enlarged Community of which Britain is a member) included these words:

"... shall encourage technical and economic research relating to the production and the development of consumption of coal and steel as well as the safety of workers in these industries. To this end it shall organise all appropriate contacts among existing research bodies."

When Britain became a full member of the enlarged Community (and, in doing so, doubled total Community coal output at a stroke), the NCB was granted full representation on the ECSC Committee of Experts with the opportunity to share in projects in mining, coal utilisation and health and safety.

In addition to the multi-national co-operation represented by the IEA and ECSC, there are numerous bi-lateral arrangements between pairs of countries which I can best illustrate by mentioning some of the contacts between the British coal industry and its counterparts in other countries. Some of these are informal meetings on specific subjects between groups of people concerned with these topics and their opposite numbers in another country while other contacts take the form of regular bi-lateral agreements.

With regard to informal contacts, I can say that in the last six years, the number of visits by NCB staff to other coal producing countries has more than doubled. Most of the visits have been to European countries but there has also been a 50 per cent increase in visits to others outside Europe where coal is important.

There has been a similar increase in incoming visits from coal industries abroad, with visitors coming from as many as 50 different countries in an average year.

Though I would not claim that all these are directly concerned with questions of research and development, many of them touch upon it in one way or another.

These contacts help to build up and maintain a healthy climate of cooperation and mutual self-help. It is against this background that the more formal international agreements are best able to flourish and expand. Collaboration and cooperation on an international scale now exist at many levels and cover a wide variety of topics.

Particularly important are our contacts with the United States with its extensive, and growing interest in coal research. In June 1974 I signed an agreement with Rogers C.B. Morton, then US Secretary of the Interior, setting up a programme for the extensive exchange of research ideas, results, programmes, staff and even equipment, covering every aspect of the coal industry, from prospecting to end use and disposal of waste. More recently, in June 1977, I signed a technical research agreement with the National Coal Association, which represents the bulk of American coal operators, and whose President, Carl Bagge, is a close friend of British coal.

We also have active research agreements, both on coal mining and utilisation, with West Germany and Poland. Our agreement with Canada is principally concerned with coal utilisation whilst those with Hungary and the Soviet Union relate to mining technology. We have recently signed a further agreement with Australia.

I should like to make a special reference to another way in which we in NCB are seeking to assist the spread of technology. This is an interesting venture in which the NCB is collaborating with major British companies to offer our joint expertise to overseas coal industries on a consultative basis. In this way, other coal industries will reap the benefits of our experience while we receive what we believe is a reasonable reward for our services, to plough back into our own coal industry.

As the biggest coal mining enterprise in the Western world, I believe we can offer a great deal to other coal industries and there is already considerable interest in the service we are offering. On our part, the NCB will be able to participate more fully in the overseas development of coal, for the ultimate benefit of the world energy economy as a whole. Our UK partners in these enterprises – Coal Processing Consultants, which is concerned with coal utilisation, and PD/NCB Consultants, which is concerned with coal utilisation, and PD/NCB Consultants, which deals with mining, are Woodall-Duckham Ltd, and Powell Duffryn Ltd. respectively. Another company, Combustion Systems Ltd., has also been set up jointly with British Petroleum and the National Research Development Corporation to exploit our expertise in fluidised bed combustion. Overseas Coal Developments (OCD) is a joint enterprise for the development of coal operations abroad and is now active in Australia, Latin America and elsewhere.

Twin-Pronged Role of Research

It is clear, then, that in every coal producing country, in greater or lesser degree, there is an increasing commitment to expenditure on coal research, coupled with an expansion in international co-operation in this field. Furthermore, there seems to be tacit agreement on the general lines which present and future programmes should follow. Although different countries place special emphasis on particular factors, depending on the local situation, there are two broad areas into which all research and development falls.

These are, first, projects designed to develop new and improved methods of prospecting for coal, extracting it from the earth and processing it, ready for dispatch to the user. This covers a wide variety of activities, from geophysical studies and investigation of environmental conditions underground to mining machinery, roof support systems and surface coal preparation plant.

Only by intensifying research and development work on mining problems of this kind shall we be able to produce economically the considerably larger tonnages of coal we shall undoubtedly need to close the forthcoming energy gap.

Secondly, there is the complementary and equally important set of problems associated with new and more efficient ways of utilising the greater quantities of coal we aim to produce. We need better ways of burning coal in existing applications – from electricity generating stations to blast furnaces – and the development of entirely new methods of using coal in order to widen the range of end-uses, and so release oil and gas for applications where, as supplies decline, they will be urgently required.

In brief, then, any coal research programme must address itself to the twin problems of mining technology and coal utilisation. This is recognised in the programmes of all national and international bodies and not least in Britain where NCB research and development operations are largely based on work at two major Establishments – one dealing with mining problems (the Mining Research and Development Establishment – MRDE – at Bretby) and the other devoted to utilisation (the Coal Research Establishment – CRE – at Stoke Orchard, near Cheltenham).

Mining Methods

Because we must mine our coal before we can put it to useful work, I propose first to give an outline of the functions and present activities of MRDE.

At Bretby, a staff of around 1,000, including a very large proportion of qualified engineers and scientists, covering the many different disciplines which are involved in modern mining, has the use of fully equipped laboratories where new machines and equipment can be tested and workshops where experimental and prototype models can be built. At nearby Swadlincote, land formerly occupied by a colliery provides a 25-acre site for testing large items of plant, such as underground conveyor systems, transport equipment and coal cutting developments, under accurately simulated real-life conditions.

Bretby's philosophy recognises that the kind of research and development on which it is engaged has four main roles.

First, it is necessary to identify problems which it believes it can help to solve, to establish the priority in which they must be tackled and maintain collaboration with line management in the industry on current progress and future plans.

Secondly, existing engineering and scientific knowledge must be applied to the solution of these problems, making use of technology from other industries where this may be applicable to coal mining. This explains why MRDE needs a staff which covers such a wide range of disciplines.

Thirdly, there must be investigation and evaluation of all natural phenomena relevant to coal mining, in order to gain a better understanding of these matters. Most of this is carried out within MRDE but progress in other research establishments is also monitored, both in the UK (for example, the Safety in Mines Research Establishment) and overseas.

Fourthly, and finally – perhaps even most importantly – it is essential to ensure that all developments are reported to those in the industry who have the opportunity to apply them. There are plenty of examples throughout every industry – I doubt whether the coal industry has been entirely guiltless in the past! – where successful developments have never been applied in practice owing to lack of communications between the research establishment and the engineers "in the field". It should be the responsibility of the research and development function to ensure that line management is aware of all developments and given every assistance to apply them at the point of production.

We believe, therefore, that it is important for any research and development facility of the scale of MRDE to base its operations on a definite philosophy. It is equally important to recognise that it is in the nature of all research and development, certainly in an industry like coal mining, to be evolutionary, rather than revolutionary. There is a pattern of step-by-step advance, with each successive development based on earlier progress, leading to machines and methods better than those from which they have been developed.

Every now and then, however, a revolutionary rather than evolutionary development emerges from the work programme, as has happened three times since the Second World War. These were the armoured flexible conveyor which led to the working of longer coal faces, the Anderton shearer-loader which gave us an almost universal method of winning coal, and powered pit supports which enabled continuous winning of coal with a dramatic reduction in manpower.

It is also important to realise that the completion of any development, evolutionary or revolutionary alike, inevitably takes time – perhaps 5-15 years for an evolutionary step and 10-20 for a radical departure. With so much at stake, and not least the lives of miners, the process cannot be hurried but must go through a series of essential stages, from the original conception of the idea, through prototypes, exhaustive pre-production model testing under all possible conditions, the making of production machines and their installation on location in sufficient numbers for their effectiveness to be established under actual real-life operating conditions.

From what I have said so far, it will be clear that the lead time to bring all research and development to fruition is so long that the necessary work to meet end-of-the-century demands for more coal must be put in hand now and the appropriate funds made available without delay, if we are to be allowed to make our optimum contribution to future energy needs.

What are the chances of a fourth revolutionary development to follow the three to which I have referred?

Naturally, every research worker worth his salt is always peering into his evolutionary workload for a sight of a possible revolutionary development. MRDE are no exception and we believe that the next major change will lie in the field of remote and automatic control, applied to every phase of the mining operation. This could be the fulfilment of previous experimental attempts in this direction, brought to fruition by the continuous evolutionary work in the past and further such work now in hand and to be continued into the future.

Our previous experience included a development known as ELSIE (Electronic Signalling and Indicating Equipment), ROLF (Remotely Operated Longwall Face) and the system planned for Bevercotes Colliery, where it had been intended that all operations throughout the pit should have been remotely or automatically controlled.

Though these experiments did not achieve all the success expected – partly because the equipment at that stage was too complex – we continued to work on solving the problems which had emerged, and to take advantage of developments in other industries, such as solid state circuitry and computer advances. These and other developments promised greater reliability and flexibility, at reduced cost.

If we can bring our latest concepts of remote control and automation to a successful conclusion and apply them to an increasing number of mining applications the impact could be dramatic for three main reasons.

First, automation will permit better plant utilisation, because it will be based on reliable and better information and because an automated, computerised system can solve problems which are either too complicated or require too much computation for the human brain to solve in the time available for effective decision taking.

Secondly, manpower can be deployed more satisfactorily, enabling machines which, thanks to research, will be more reliable, to be remotely operated from a central point.

Thirdly, management will be able to make better decisions because the new systems will be providing more and better information.

Although the groundwork for the necessary technology has been laid with the experience with ELSIE, ROLF and Bevercotes, and in some ten years of continuous experiments since then, much remains to be done. For example, it is necessary to establish the reliability of all the component parts of the system and many of these are complex items in their own right, as well as the integrity of the system as a whole.

We must then develop devices for measuring and transmitting all the variables in the system and for analysing all the information obtained. Finally, control systems must be developed which will use the information supplied to detect variations from the pre-determined programme and initiate the necessary corrective action automatically. All these stages must be applied to all mine operations as an integrated overall control system.

Let us now see how all these aspects of the future remote and automatic control of mining operations are accommodated in the current programme of work at MRDE.

Programme of Work

Although MRDE's Director, Peter Tregelles and his staff have this next major development prominently in mind as they carry out their research programme, it is not solely directed towards the big breakthrough but provides the necessary evolutionary background out of which mining automation will, hopefully, emerge, together with other smaller, but useful developments.

A recent re-arrangement of the work programmes has been designed to facilitate this dual objective. Instead of projects being grouped in accordance with different scientific disciplines as formerly, they are now allocated under headings that are more closely related to the various colliery operations. There are seven groups, namely, coalface, roadway drivage, underground transport, coal preparation, comprehensive monitoring, basic studies and testing. Although no specific reference is made to safety in the list, I should emphasise that safety is an over-riding consideration in all MRDE's work. No policies are ever implemented or any decision made without a thorough examination of all safety aspects. This applies all along the line, from research into the impact of dust, methane and other environmental factors to the development of mining systems.

A closer look at the seven headings in turn will show what is involved in each area and, collectively, how they are designed to advance the productivity of Britain's coal mines. This is not the place to refer to technical detail but rather to try to pinpoint the implications of the various developments.

Under the heading of coal face operations, there are four work groups, each comprising a number of projects – a general pattern which also applies to the other six headings.

One of these work groups is concerned with the reliability of coal face equipment, for example, by carrying out life tests on alternative types of machines, while another group is seeking ways of raising productivity by such means as obtaining more coal from each "pass" of the coal cutting machine. A third work group is looking for improvements in the ancillary equipment – pumps for example – while the fourth is working towards automatic control in this area, for instance, by developing automatic steering systems for coal shearing machinery.

Already we see how the search for remote control and automation ("advanced technology mining" or ATM as we now describe it) is inherent in all Bretby's work.

Under roadway drivage the four work groups – in-seam drivage, mixed coal and stone drivage, hard rock drivage and ancillary operations – are concerned with such problems as the more rapid driving of roadways under widely varying conditions of seam thickness and soil and rock type and the evaluation of different machines, either developed at Bretby or in collaboration with mining equipment manufacturers. Incidentally, this cooperation is a permanent and valuable aspect of MRDE's activities. Remote operation figures prominently in this work group too, for example, by controlling machines by radio.

Three work groups make up the heading underground transport, respectively dealing with the optimisation of belt conveyor systems, mineral transport by pipeline and the transport of men and materials. These are largely self-explanatory topics which also include a strong element of ATM, as, for instance, computer control of conveyor running. We have a practical example of this in the form of a prototype system at Bagworth Colliery where a mini-computer operates a closed loop conveyor control system which is already having a good effect on conveyor performance. Interesting, too, is the movement of coal through pipeline, either pneumatically or hydraulically and trials are being undertaken at two collieries.

Because the basic processes of coal preparation are well established, the work in this sector – the groups deal with dry fines extraction, fine coal and tailings, reliability of products and processes, and automation and control – is largely directed towards the monitoring and control functions needed to improve the quality and consistency of the end product and the reliability of plant and systems. This work is clearly leading towards automatic control of an important sector of mining operations.

When we come to consider the work groups listed under the heading of comprehensive monitoring – the development of sensors, data transmission systems and computer control – we are dealing with work which lies at the heart of ATM and which therefore pervades all the other work groups. The work begins with the detection and measurement of the variables involved in machine and system control and the transmission of this data.

Mini-computers, with their capability of processing this data into visual displays and printouts, are already demonstrating the vital role they will play, as shown at Bagworth, the first of many such prototype installations. Radio systems are being introduced as a means for controlling machines in mobile situations and automatic techniques are also being applied to the important task of monitoring methane concentrations and ventilation levels, with an actual system under trials at Brodsworth Colliery.

Basic studies is a field of work covering work on phenomena and materials which are peculiar to our industry. The first of the five work groups deals with the environmental problems below ground, such as dust, heat and noise while another is concerned with the behaviour of the earth's strata, one of the most important geological factors for the underground mining engineer. A third group deals with engineering principles and materials, with special attention to the principles of coal and rock cutting, metal corrosion and fatigue, and a fourth is designed to help mine managers to know more about geological conditions both at the point where they are currently cutting coal and also to know something about seam conditions further ahead of the coalface. A miscellaneous work group includes such items as soil mechanics and the use of liquid nitrogen as a power source.

Finally, MRDE's seventh heading – testing – comprises the four aspects of all successful testing work, with work groups covering approvals and specification testing, assessment of proprietary equipment and materials, investigation of failures in service and development testing.

All are concerned with the reliability of equipment and, in the process, they throw up a great deal of information which is of use to the designer. Briefly, the approvals and specification testing group examines equipment and products of all kinds, from roof supports and cutting tools to conveyor belts and lubricants, to see if they meet NCB requirements while assessment of manufacturers' equipment is made at the Swadlincote site, where mining systems are also tested. Testing of complete systems is a relatively new departure and the facilities at Swadlincote enable a number of items of machinery to be tested, under accurately simulated conditions, while working as part of a complete mining system, to ensure that each separate item is compatible with the others.

Mining of the Future

All technologies change and develop, and this is true of coal extraction no less than of any other industrial undertaking. It is a wise industry that looks to the future and plans accordingly. The NCB recently (1976) commissioned a small team from amongst its scientists and engineers to try to perform the seer's role, and visualise what might be the pattern of mining into the twenty-first century. There are many pitfalls to this sort of clairvoyance but many benefits as well: if the forecasting is shrewdly based the technical problems that still need to be solved can be tackled in good time and the appropriate research and development instituted.

The team took the view, based upon innumerable examples in the history of technology, that revolutionary developments in a specific industry would be more likely to have their origin outside the industry than in it. The new findings could at the moment be the academic curiosities in a field of research which is quite unrelated to coal mining. The team therefore decided to visit the country's main research establishments, in both the public and the private sectors, to find out what was going on, and what was exciting engineers and scientists. Altogether some 30 establishments, all of whom co-operated cordially in the exercise, were visited.

There are two major alternative possibilities for coal extraction in the next century. One is that the mineral should be brought out in solid form as at the moment, the other that the latent energy of the coal should be transformed in situ so that the agent of practical use comes in liquid or gaseous form. In the matter of mechanical mining we would hope to have more efficient and more reliable cutting machines, with a solution to the *present bete* noire: the difficulty in cutting very hard and very abrasive rocks. We look forward to the development of new hard cutting materials by applying the discipline of fracture mechanics, now a very live topic in materials science. As a supplement to mechanical cutting there might be cutting by lasers, though this would pose environmental problems, and more optimistically, by water jets at very high pressure.

Many engineers have been fascinated by robot devices. A possible role for robotry might be the use of 'telechiric' devices, which would perform quasi-manual actions under the control of an operator a short distance away. 'Telechirs' could do jobs and perform manipulations under circumstances that might spell danger for human operators.

The NCB team listed about a dozen possible techniques for 'in-situ energy extraction' ranging from underground gasification, which has so far had limited practical application, to others such as pyrolysis or solvent digestion which are based upon no more than laboratory experiments or even purely theoretical speculation. These will be the techniques to use if a climate of opinion were to arise in the twenty-first century which would preclude men from going underground. It is important that the technical and economic possibilities of these methods should be thoroughly aired in good time, so that the choice for major investment should be wisely made.

An interesting suggestion is the use of bacteria to break down coal into simple organic compounds. This ought not to be exaggerated as an ultimate practical possibility, but basic research along these lines might pay off in unsuspected directions.

Perhaps the most fascinating of the visits made by the NCB team were to the laboratories of the big electronics firms. The compactness, reliability and versatility of solid state instrumentation will create the possibilities for a high degree of automation in mining, whether the process be extraction of mineral or provision of its alternative. Developments in radio and telephone communication offer the prospect of trouble-free speech transmission. There are also major developments in computer technology which will be relevant to mining. Data storage, retrieval and analysis will have wider and wider scope. Data displays will be more various, selective and dramatic than at the moment, providing management with vastly improved supervision and control.

While the exercise of looking to the future did not, end with the emergence of anything particularly unexpected, much was seen that is still being mulled over and which will guide research until the end of the century.

Theory Into Practice

I have already referred to the need to ensure that Bretby's developments – to which I have referred only in general terms above – are translated from research and development stage into full production – a process which can take far too long, even if some promising developments are not lost along the way.

At MRDE, we have taken steps to reduce development lead times and make sure that projects with a good chance of success move smoothly through all stages.

We have established a National Mining Research and Development Committee, under the chairmanship of John Mills (NCB Mining Board Member) and whose members represent, at high level, the research, mining, planning and training sides of the industry, plus the Director of the Safety in Mines Research Establishment. This influential Committee gives directions on matters of policy and major investment, monitors progress and keeps a close eye on rates at which developments pass into practice.

So much for overall direction and encouragement of research and development. The more detailed supervision of research under the seven headings I have described is in the hands of Major Development Committees. There are five of these, one each for the five operation fields of coalface, roadway drivage, underground transport, coal preparation and comprehensive monitoring. These committees advise on plans, comment on progress, assist in the provision of test facilities at collieries and promote the exploitation of successful projects. They thus provide an effective bridge between development and exploitation but also encourage other developments, for example, by NCB areas and manufacturers.

As to the two remaining headings into which Bretby's work is divided, the work in basic research is covered by the Research Liaison Committee which sees that the work matches the industry's long term needs, while the Testing Committee endorses the programme of work and sees that any lessons learned are not thrown away.

I want to stress the important role of the Major Development Committees which, after deciding on an initial development, then sponsor the pre-production model by finding suitable testing sites where on-site experience can be obtained. The Committees also encourage developments from elsewhere in the UK and overseas, including finding test locations, so that MRDE does not duplicate work being carried out by others.

Another measure which has been introduced to speed the transition from theory into practice is the setting up of a Project Development Division which will take over projects which have reached the prototype stage, carry them through pre-production proving in a variety of situations and prepare them for exploitation.

There remains one more aspect of a successful research and development operation – the problem of training the staff who will have to operate the new equipment and systems which make the grade. MRDE has always been involved in training operators to instal, operate and maintain new equipment, using a variety of training methods, from "on the job" instruction to formal courses.

As the new technology emerges, the need for training increases and, to cope with the need for "technology transfer", a training manager has been appointed at Bretby where he and his staff will be accommodated in a new Training Centre. Training schemes, for NCB staff at all levels, will ensure that the developments which enter the pits and other NCB installations in the future will be used to best advantage by well trained personnel.

I hope I have said enough to demonstrate that the Mining Research and Development Establishment has the facilities, the policies, the projects and the personnel to enable the industry to produce the coal the country will need to close the energy gap ahead. Assuming that this can be done at reasonable cost – and I have no doubts on that score – there is a complementary need for research and development into ways for making the best use of the larger quantities of coal made available by MRDE's work.

Putting Coal to Work

To carry out the task of investigating and developing new ways of using coal and, in particular, converting it to other forms of energy and materials, we have the NCB Coal Research Establishment (CRE) at Stoke Orchard, near Cheltenham.

As at Bretby, the research teams at Stoke Orchard are working on major long-term problems which fall into one of three different categories. One category may lead to processes which have almost immediate commercial applications, others may take a little longer to materialise, say five years, while some are aimed at anticipated requirements which lie perhaps 15-20 years ahead.

This allocation of projects between these three time scales gives only a rough and ready guide to CRE's work, which is on evolutionary lines, so that new and improved products may appear at any time. The paths of research can take sudden and unexpected turns. A long-term project may be concluded earlier than expected while, on the other hand, an immediate development may suggest further lines of research leading to a long-term project.

An example of CRE's work in the first category – that which is available almost immediately – is our work on metallurgical coke. The iron and steel industry is a traditional market for coal, with a requirement of around 20 million tonnes a year, and looks set to remain at this level.

It is therefore important for coal to retain this market but by no means all coal is equally suitable for turning into the coke which the steel industry needs. The best coals for this purpose come from South Wales and Durham but supplies from these areas are becoming scarcer and therefore likely to grow more expensive.

Although Britain's steel industry, and others around the world, have for some time been investigating alternative methods of iron production, such as direct reduction, which do not need coke, it is clear to me that the conventional blast furnace, dependent on coke, will continue to be the most widely used method of iron production for the foreseeable future.

So, faced with the assurance of a continuing demand for metallurgical coke over a period when supplies of prime coking coals are diminishing, the research workers at CRE have already achieved considerable success with their investigations into techniques for blending inferior types of coking coal, not excluding some which would not normally be classed as coking coal at all, to produce coke which gives good results in blast furnaces and cupolas. In all this work, we are collaborating closely with the British Steel Corporation and the British Carbonization Research Association (BCRA).

To give an indication of the progress that has been made, compared with, say, twenty years ago when only prime coking coal was supplied to the coke ovens, today, no more than about 25 per cent of the coal so used is of this type.

I would not like to give the impression that it is just a matter of trying out different types of coal in varying proportions in the hope of arriving at the best "mix". On the contrary the work at CRE has established a scientific basis for blending coals to produce coke of any desired quality and so to make the maximum use of those coals which comprise the bulk of the reserves of the United Kingdom. The work has involved thoroughly understanding the carbonisation process and has shown the importance of efficient blend preparation. Already high quality coke for metallurgical use is being manufactured in coke ovens in the North East and Midlands from blends not containing any prime coking coal. With the new techniques being introduced into the coking industry, such as heating blends prior to charging them into coke ovens, we envisage a widening of the range of coals that can be used. The need now is for guidance from blast furnace operators on the properties of coke required for the blast furnaces of large diameter and high capacity now being built in the United Kingdom.

Another approach to the problem is to make briquettes of weakly caking or non-caking coal and feed these to the blast furnace. This, again, is not such a simple matter as might appear but the work in hand at CRE for turning low ranking coking coals into briquettes (or formed coke) is showing promising results. Blast furnace trials so far conducted by British Steel have been encouraging and one possibility is that the use of formed coke briquettes, in a blend with conventional coke, may prove worthwhile.

Before I leave CRE's work on metallurgical fuel I should like once more to stress its importance in the context both of Britain's and the world's future energy problems. By successfully using lower grade coals in steelmaking it will lower the costs of one of Britain's basic industries while, by eliminating the need to import coking coal it will reduce the demand on a world resource which is also diminishing, releasing supplies for countries which are less well placed than ourselves.

Coal and Electricity

Coal's largest single market is in electricity generation, which takes at present about 70 million tonnes a year, competing with nuclear power, oil and, to a limited extent, with gas. Anything we can do to increase the usage of coal in this bulk heating market will release corresponding quantities of oil and gas for the premium uses for which they are essential and for which the relatively limited supplies should be reserved.

This explains the wide interest, both in the UK and elsewhere, in new techniques of coal usage in electricity generation. Undoubtedly, the most important development in this field is the technique known as fluidised bed combustion, described by Dr. J. Gibson, now NCB Member for Science, as "the only unique post-war combustion development." Important though it is for electricity generation, fluidised bed combustion has much wider future applications throughout the industrial and commercial heating market.

To explain briefly, fluidised bed combustion is a technique in which coal is burned in a bed of finely divided particles of any suitable mineral matter, such as ash from previously burned coal, limestone or sand. Air is passed upwards through this bed so that the particles become highly turbulent in the air stream and mix rapidly. In this condition they behave rather like a liquid and hence the description fluidised bed. Operating temperatures of the bed lie between 750°C and 950°C so that tubes passing through it can be used to produce steam rather like the tubes in a conventional boiler.

This is a layman's description of a technique which was first successfully demonstrated in Britain during the mid-1950's. Preliminary work was done by the Central Electricity Generating Board but because of CEGB's heavy involvement in other developments at the time, including nuclear power, the subsequent work on fluidised combustion was taken over by the NCB.

A potted history of the development of fluidised bed combustion since then illustrates the problems of any major research project, however promising.

As long ago as 1969, several small experimental rigs were in operation, including a 4.5 m (14 ft. 9 in.) high combustor at CRE, which gave promising results. CRE then built a larger combustor, as a section of a water tube boiler, as used for power generation, with a heat output of 0.5 MW.

NCB's involvement continued with work at the British Coal Utilisation Research Association (BCURA) Laboratories at Leatherhead (now NCB's Leatherhead Laboratories), where two fluidised bed units were constructed – a 2.5 MW vertical shell boiler for industrial heating applications and a 2 MW combustor for electrical power generation. The CRE and BCURA programmes, together with back-up research, added up to over 22,000 hours of research and development.

All this theoretical and practical work brought us to the point where, by 1971, we felt confident we could design a 20 MW demonstration fluidised bed steam boiler but, as Government funds were not available for such a project at that time, the development programme was interrupted.

Meanwhile, in the United States, the Office of Coal Research (OCA) and the Environment Protection Agency (EPA) had been sponsoring work on fluidised bed combustion and the EPA had jointly financed, with the NCB, some of the work carried out at CRE and BCURA. When the programme came to an end in 1971, the Office of Coal Research took over complete support for the high pressure side of this work.

Here we have an encouraging example of the international cooperation in research to which I referred earlier in this chapter.

This meant that the work on fluidised bed combustion had not been wasted and, still further to make use of NCB's expertise, the National Research Development Corporation (NRDC) took over its exploitation commercially. Then in 1974 a joint company – Combustion Systems Ltd., as briefly mentioned earlier – was formed by NCB and NRDC, together with British Petroleum. In collaboration with this new company, the well-known boiler manufacturers Babcock and Wilcox successfully converted a 13.5 MW water tube boiler to fluidised bed combustion. This led to the formation of another joint company – Babcock-CSL – to ensure that not only our expertise would be available overseas but British fluidised bed equipment also.

This is by no means the end of the story but, before bringing it right up to date, let me outline the advantages of this very important development.

Fluidised bed combustion has many benefits. The combustion of coal under this system is not affected by its type, ash content or moisture content. Low grade coals which have not hitherto been considered suitable for combustion because of their high or variable ash content can be used efficiently in the fluidised bed. This effectively increases the quantity of our coal reserves which can be economically used, compared with present methods.

Because fluidised bed boilers are more compact than conventional equipment, it will be possible to achieve lower unit capital costs and shorter construction lead times. Recent design studies have indicated that capital cost savings for a full scale power station based on the fluidised bed could be 14 per cent for an atmospheric pressure design and as much as 23 per cent for a pressurised unit, compared with the conventional thermal station operating on pulverised fuel.

These are very important advantages but there are others, including the contribution fluidised bed combustion can make to a better environment. The emission of sulphur dioxide which occurs when fuels containing sulphur are burnt can be reduced with the fluidised bed simply by adding limestone or dolomite to it.

Other benefits arise from the low combustion temperature of the fluidised bed. For example, it reduces the formation of nitrogen from atmospheric nitrogen and although the extent of oxide formation depends on the way the fluidised bed is operated, emissions are in all cases lower than with conventional boiler practice. The low combustion temperature reduces the vaporisation of alkali salts from the ash, minimising fouling of the boiler tubes, caused by condensation of these salts. It also reduces their emission to the atmosphere.

Although fluidised bed combustion has been primarily considered for electricity generation, it can obviously be applied in many other directions, such as hot water boilers, horizontal shell-type steam-raising boilers and packaged shell and water tube types. All these are widely used in industry where they are generally oil or gas fired at the moment.

It is also possible to use the system to produce hot gas for industrial and agricultural drying processes, such as the first commercially operated fluidised bed in operation at Widnes where it produces hot gas for drying grass to make cattle feed. In this case the hot gas direct from the bed is diluted with air to the right temperature but an indirect system, using tubes in the bed, could be used where a clean hot gas is needed.

Fluidised bed combustion techniques are also ideal for incinerating waste materials, recovering the heat in many cases. At CRE an experimental plant has been developed to incinerate the waste slurry from coal preparation plants. The slurry is concentrated to about 50 per cent water content and sprayed onto a fluidised combustion bed. The technique is being applied to other industrial slurries, sludges and wastes, including the incineration of sewage sludge.

All these developments take advantage of the inherent characteristics of fluidised combustion which is so efficient that there is no emission of unpleasant compounds. Moreover, CRE experience confirms that many materials not normally classified as fuels can be successfully burnt in fluidised beds.

One more exciting possibility arising out of the development of fluidised bed combustion can be mentioned and that is the coal fired gas turbine. Previous attempts with coal or fuel oil have failed due to rapid corrosion or erosion of the turbine blades, but the reduced emission of alkali salts and vanadium at the low combustion temperature of the fluidised bed reduces the risk of blade damage. Moreover, the ash from fluidised bed combustion is not abrasive and any ultra fine particles which pass through the turbine's gas cleaning system do not erode the blades.

While Britain still holds a technical lead in fluidised bed combustion our future work in this field is closely bound up with the international effort which, in addition to development in the United States and Western Germany, is taking place under the aegis of the International Energy Agency. An experimental facility is planned at Grimethorpe Colliery by the IEA's Coal Working Group, under the joint sponsorship of the United States, Germany and the UK.

Covering a 7-year programme, the Grimethorpe facility is being designed in such a way that it can be operated under a wide range of different conditions. This will enable it to be used in the interpretation and correlation of data which become available from the development projects in the three countries, for the benefit of the three participants and any other IEA member countries who may wish to join the project.

I sum up the fluidised combustion story as follows:

Ten years research and development, in which we in the NCB can claim the pioneering role, has resulted in the building of prototype units for a number of applications. The next ten years will give it increasing importance, both in manufacturing industry, where it will offer the opportunity to convert from oil or gas to coal, and in power generation where it will permit lower quality coals to be used and so extend the amount of exploitable coal reserves.

Looking Ahead

Into CRE's third category of research projects – those which are being actively pursued at the present time but which may not be applied commercially for many years – fall two lines of research, the aim of which is to produce from coal a wider range of energy products. These coal conversion developments are based respectively on liquefaction and gasification.

In the case of liquefaction, there are two main processes – by synthesis, when a synthesis gas is produced from coal and steam and then converted to the type of feedstock required by subsequent treatment, or by solvent extraction of coal and hydrogenation.

Synthesis techniques have been exploited for some years, notably in Germany before and during the Second World War and currently in South Africa, but we believe that solvent extraction will eventually prove to be the most economic route for feedstock production. In this process, coal is first dissolved in oily solvents, and, after filtration to remove mineral matter and undissolved coal, the solvent is removed and the residue hydrogenated to produce the types of feedstock required, using hydrogen also made from coal by reaction with steam.

CRE is working on the solvent route, using coal tar oils as solvents. The research team believes that they have solved the practical obstacles of the process, certainly on a small scale and there should be no difficulty in scaling up. The advantages of producing liquid fuel from coal in this way are twofold. First, the crude product aligns closely with the characteristics of crude oil and so its refining can be carried out using known and proved petroleum technology. Second, the prospect of having a continued availability of familiar liquid fuels in this way means that the need to develop new transport technology (such as the electric car) is lessened in the short term and research effort can instead be directed towards improving the performance and characteristics of the internal combustion engine. I was particularly pleased to be driven around in a car powered by solvent refined coal fuel at the Mining Festival in Blackpool in November 1977 which I hope was an illustration of things to come in this field.

Turning to possible end products, coal extract can be used, without hydrogenation, to make very valuable forms of carbon, such as carbon fibre and electrode coke. The production of carbon fibre from coal has been successfully demonstrated at the laboratory stage but, at the time of writing, the process would not be economically viable. Its day will come, however.

On the other hand, we feel that the production of electrode carbon, for use as electrodes in arc furnaces in the steel industry or in aluminium production, may be economic even now. Test electrodes have been produced and, put on trial by the British Steel Corporation in a small arc furnace. Their performance was entirely satisfactory, when compared with commercial electrodes currently available.

In another technique of solvent extraction, which CRE has pioneered, coal is treated with hot compressed gas. The gas readily penetrates the coal, some of which goes into solution in the gas phase, leaving a char which contains the mineral matter. If the gas phase is then transferred to another vessel at a lower temperature or pressure, the solvent effect disappears, the coal extract is precipitated and the gas is free to be used again. The coal extract is thus recovered, without filtration, in an ash-free condition.

Compared with the use of liquid solvents, this system has a number of advantages including separation of the solution from insoluble matter without the rather cumbersome filtration necessary with liquid solvents, the easier separation and recovery of the gaseous solvent and the production of gas extracts with lower molecular weight and richer in oxygen, making them more suitable for the eventual production of hydrocarbon oils and chemicals.

Coal into Gas

It may appear rather strange for the coal industry to be concerned about turning coal into gas at a time when North Sea supplies of natural gas are still relatively abundant and it is true that at the present state of our work on coal gasification and under present market conditions, the processes would not be economic. However, as I have shown, supplies of North Sea gas will begin to decline, possibly by the end of the century, and we shall be looking to coal as the source of substitute natural gas (SNG).

Even now there is an impending shortage of natural gas in the United States, where the production of SNG from coal is of considerable current interest. At the very least, therefore, we feel it is also in the UK national interest to be aware of what is going on internationally and to be involved in it to the extent that we have the know-how in readiness for the time when it will be needed here.

Meanwhile, we are looking to the gasification of coal for the production of a lean gas (with a calorific value of about 150 Btu compared with natural gas at about 1000 Btu) as a clean fuel for gas turbines. These would be used in power generation by advanced systems incorporating combined gas and steam turbines – systems which would be of higher efficiency than the simple cycle of power generation currently in use. It is expected that developments in gas turbines, enabling them to operate at higher inlet temperatures, would lead to even higher generating efficiencies.

Experience gained during this research into coal gasification would be invaluable if at a later date, as seems likely, it becomes necessary to investigate the manufacture of gas for other purposes, such as SNG for general consumption, reducing gas or even synthesis gas.

At the NCB Leatherhead laboratories, work is taking place on pyrolysis processes. In this work, the volatile materials in coal are removed by suitable thermal treatment, leaving a char residue. The volatile components distilled from the coal can be utilised as feedstocks for the chemical industry while the char product can either be used for combustion or, alternatively, gasified to a synthesis gas for eventual conversion to SNG.

Pyrolysis is thus yet another of the many options open to Britain's coal industry, in readiness for the day when we need to close the energy gap. It will be too late to undertake complex research projects when the moment of truth is actually upon us. Hence the importance we attach to the work now in hand in our laboratories and in the field.

There is a further step which will become increasingly important, namely that of bringing together the various conversion processes into a single plant. Thus, a coal refinery or complex is conceivable in which coal-based processes producing a range of products, such as electricity, substitute natural gas, oils and chemicals, are combined together in such a way that the overall capital costs of the processes are reduced and the overall efficiency increased. This can be achieved by using rejected heat or by-products from one unit process in another. The system should also be flexible enough to meet fluctuating market demands.

Everyone Benefits

Any research which helps to increase the national output of energy at economic cost for a longer period ahead than would otherwise be possible benefits everyone and this is the basic aim of all NCB research, whether in the field of mining problems or of utilisation. Though I have perhaps tended to concentrate on the two big customers – the steel industry and electricity generation – we do not neglect the many other consumers of coal, in industry, commerce or the home.

On the industrial and commercial fronts, the work at CRE includes forward-looking work on the application of fluidised bed combustion to the smaller-scale requirements of industry, as already mentioned.

An example is the shallow bed technique which has been developed for the small to medium size boiler, and which is believed to offer a number of advantages. It can be used with a wide selection of coal types and sizes (in size from 0 to 1½ in and with ash content up to 30 per cent). Low operating temperatures mean less wear and tear on the boiler with consequent lower maintenance costs. High burning rates can be achieved, especially if a new technique of artificial bed cooling is applied, and there are the lower pollution rates associated with fluidised bed combustion.

This is for the future. In the meantime, work continues on conventional burning plant for industry, with further studies on automatic stoking and de-ashing. Close links are maintained with the manufacturers of boiler plant whose research and development departments are working on the same problems as well as on other aspects of more efficient and economic boiler operation.

Another section of CRE deals with the appliances which are supplied to the 5 million domestic solid fuel users in the UK. The Establishment's work covers the development of new types of solid fuel appliances and the further exploitation of existing types. Its major achievement in recent years has been successful research on domestic appliances burning bituminous coal smokelessly. A number of these appliances are now being marketed by leading manufacturers.

Close links are maintained with solid fuel appliance manufacturers and the results are seen in the wider selection of equipment now on the market, including new types of open fire, combined open fire/back boilers and various kinds of roomheaters, such as the popular glass-fronted type and the more recent "hole-in-the-wall" unit.

All this work will find its justification in the homes of the industry's domestic consumers, just as our other research and development work, both at Stoke Orchard and Bretby, is justified by the contribution the coal industry is making to the national fuel economy – a contribution which, I am sure, can only grow as the years pass and the reserves of other fossil fuels dwindle.

Research on Other Fuels

I have written at some length on coal research and development, partly because of my special connection with the subject but also because I believe that improved methods of coal mining and utilisation will have an enormous influence on the long-term energy prospects of Britain and the world. There is, of course, an equal need for research and development in other energy fields and I shall end this chapter with a broad outline of the research scene in these other fields as I see it.

Work on a national basis is carried out by British Gas, by the Central Electricity Generating Board and by the Atomic Energy Authority, each of which maintains research establishments covering various aspects of their operations. The oil companies have their own research and development arrangements as do the manufacturers of equipment used in the production and utilisation of energy. Various research associations and establishments, both government sponsored and industry-based, also carry out energy-related research.

All this research and development resembles our work in the coal industry in that it falls into the two same main categories – better ways of producing the energy and improved methods of utilisation. The emphasis varies from one fuel to another.

British Gas, for example, no longer has to manufacture its fuel as in the past when "town" gas was produced, first from coal and later from oil, but now obtains it as a naturally occurring fuel from the North Sea. There remains the problem of getting it from the well-head to the shore terminal and processing it there prior to distributing it through the comprehensive grid system.

This has entailed a great deal of research at the British Gas Engineering Research Station, Killingworth, Newcastle-upon-Tyne, into a wide range of problems, including materials and construction methods for high pressure pipelines. Research continues, both on gas transmission and on many other engineering problems such as ways of storing gas in order to even out the considerable peaks and troughs of gas usage.

With so much of the nation's present fuel supplies depending on the integrity of a pipeline system which is transmitting more gas than ever before at much higher pressures, special attention is being paid to the testing of pipes prior to installation and to the development of techniques for the internal examination of high pressure pipelines while still in operation.

Much work is also being done on methods for leakage detection and repair of the distribution mains and services which carry the gas at lower pressures to the consumer. New techniques and equipment are under development and evaluation, aimed at ensuring continuity of supply with minimum environmental inconvenience.

Dependent as it is on a finite supply of natural gas, the industry is necessarily interested in methods for producing SNG from suitable feedstocks, both to provide for its future when North Sea natural gas is exhausted, and, in the meantime, as a peak-lopping facility during temporary peaks of demand. Existing techniques based on petroleum feedstocks are available, but since oil is unlikely to be available, at reasonable prices, much longer than natural gas, the gas industry is looking to coal as a source of future SNG.

For many years the Gas Industry has been engaged in research, much of it on a substantial scale, into the gasification of all potential feedstocks, including coal and into the purification of the products. At first the objective was the production of town gas but since the mid 1960's the same basic process steps have been recast to yield as the product a substitute natural gas (SNG) wholly interchangeable with North Sea gas. Much of this work was conducted at the Midlands Research Station of British Gas but some years ago urgent requirements for process information on coal gasification, especially in the US, led British Gas to establish at Westfield, Fife, a Coal Gasification Development Centre. Westfield, a former "town gas" production plant, had a Lurgi gasifier in working order and main tasks were firstly, to demonstrate the conditions under which American coals could be used, widening the previous limits on coal size and type, secondly, to demonstrate on a large scale the subsequent purification and methanation steps (much more extensive than the gasification) required to produce SNG, and thirdly to demonstrate the acceptability of the product by supplying it through the local distribution system to customers in direct replacement of natural gas.

Subsequently, a further BG development, the Slagging Gasifier, has been built and demonstrated on a similar scale for a US consortium. Like the Lurgi, this is a fixed bed gasifier – i.e. the coal is fed into the top and oxygen and steam are blown into the bottom. In the Lurgi, ash is removed from the bottom but in the Slagging Gasifier the bed is operated at a higher temperature, giving increased efficiency, and the ash is removed in the molten state. This process is one of two selected by ERDA for full scale demonstration plants in the US.

Further work is in hand on more advanced processes, the main aims of which are to decrease the energy consumed in the processes and to widen the acceptable range of coals, especially to permit the fines to be used.

As a matter of interest, the other process chosen by ERDA is the Cogas pyrolysis process; some of the key experiments were carried out under contract at the NCB laboratories at Leatherhead. Cogas pyrolysis produces liquids as well as gas and such processes may have important economic advantages in the long run. These experiments have achieved considerable success and have led to substantial overseas research contracts, especially with the United States.

On the utilisation side, British Gas carries out research and development in the Corporation's research establishments and in the technical departments of the twelve Regions.

Industrial research and development are also centred on the Midlands Research Station at Solihull which thus has the two-fold task of developing new gas production processes and of increasing the efficiency of industrial gas utilisation through research into new burners, furnaces and other heating plant, as well as work on more efficient gas control systems.

Research and development on behalf of domestic, catering and commercial customers is the responsibility of Watson House, Fulham, another part of the British Gas Research and Development division where, in addition to work on heating, cooking, installation engineering, burner design, combustion problems, standards and codes of practice, British Standard tests for appliance safety and the Corporation's own approvals procedures are carried out.

The fourth research establishment, the London Research Station is also located at Fulham and acts in the role of the corporate laboratory for British Gas. Its particular fields of expertise lie in chemical and physical analysis, mathematics and computing and in biological studies. These all contribute to the achievements of the more functionally oriented research stations.

Electricity

Research and development in the electrical industry covers the generation, transmission and storage of electricity as well as better utilisation by industrial, commercial and domestic consumers.

Low power station efficiency is a universal problem and although improvements have been achieved and will continue to be achieved through the greater availability of large modern generating units, average efficiencies remain a cause for concern. It is obvious that a major improvement must await an entirely new approach which could be the introduction of fluidised bed combustion or coal gasification in a combined cycle with a gas turbine.

Power is also lost in the process of getting electricity from the power station to the consumer. In total this is about 8 per cent of which the majority is lost during distribution – the transport of power from the bulk supply point or sub-station to the consumer.

Electricity is not easily stored in quantities large enough to even out the variations in demand but various techniques are under development, including storage in large capacitors, magnetic storage in low temperature super-conducting coils and electro-chemical storage. A storage technique which has been successfully developed is the so called "pumped water storage" system where electrical energy is stored as potential hydro-electric energy. Such stations have been in operation in the UK and overseas for many years and the CEGB is currently building an 1500 MW station of this type at Dinorwic.

Utilisation of electricity is closely related to the economics of the fuel in the various applications. Though it has a captive market in some directions, of which lighting, motive power in many cases and high grade steel production are good examples, it faces strong competition on price in many other sectors.

Research is concentrated in a number of these areas, however, usually in close cooperation with appropriate appliance manufacturers. Home heating is an example where owing to there being no need for flues or combustion in its use, it offers the highest efficiency of utilisation. Much attention has been paid to the development of equipment to make use of off-peak electricity, especially the introduction of control systems to introduce a greater degree of flexibility into these systems. Also, a considerable research effort has gone into the development of heat pumps which offer the possibility of supplying heat from electricity at an effective efficiency of around 300 per cent so recovering the losses of heat resulting from its conversion into electricity at the power station.

In the commercial field, useful work has been done on the integrated environment concept for large buildings, in which all aspects of the internal environment – heating, cooling, lighting and building design are integrated into a low energy scheme with heat recovery.

Industry, excluding the iron and steel industry now derives about the same amount of its energy needs from electricity as it does from coal, although traditionally, electricity's usage has been confined to lighting and power and a relatively few premium uses. Recently, however, the Electricity Council has paid increasing attention to this sector of the market, where development of suitable appliances could, it believes, find more openings in a range of process applications, from metal finishing and vat heating to welding and heat treatment.

Development of new industrial techniques and equipment will be based on the versatility of the heating methods which electricity offers. Apart from its ability to transfer heat to a product surface from an external source (by indirect resistance heating) which is the conventional method of heat transfer employed by other fuels, electricity can produce heat in a product by passing a current through it (direct resistance heating), by electro-magnetic induction, (by direct arc or arc plasma), by dielectric loss, by microwave absorption or by the direct use of electrical discharge phenomena occurring in vacuo and by lasers.

Nuclear Research

Britain's impressive programme of nuclear research and development has placed it amongst the world leaders in this field.

Following the successful introduction of the early Magnox stations into the national electricity grid, nuclear research centred on the Advanced Gas-Cooled Reactor (the five stations in the current programme are of this type), the Steam Generating Heavy Water Reactor (SGHWR) and the Fast Reactor, with some collaborative international programmes on high temperature reactor design. The choice of the systems for introduction in the 1980's is currently under review at the request of the Secretary of State for Energy and the ultimate choice will determine the direction of future UK nuclear research.

Nuclear research can be divided into two basic types – laboratory work in the various Establishments of the Atomic Energy Authority (AEA) and practical experience at research and experimental reactors, some of which deliver power to the national electricity grid. The three reactor types mentioned above as the primary targets for recent research have been operated in this way for a considerable time.

At Windscale, an advanced gas-cooled reactor (WAGR) started up in 1962, has provided the opportunity for studying this system, developing and proving fuels and other materials and components, as well as giving operational experience of power production with AGR's.

Since 1967, the Winfrith SGHWR has enabled AEA to gain experience in operating this system, testing fuels, fuel materials and experimental development work in support of any future commercial reactor of this type.

A small fast reactor at Dounreay, the DFR, was closed down on March 23rd 1977 after 18 years of successful operation when it had completed its role in the development of the technology of fast reactors and as an irradiation testing reactor for advanced fuels and reactor components. A larger prototype fast reactor which incorporates several components of the right size for a commercial scale reactor including fuel elements, in operation at Dounreay since 1974, and supplying power to the grid regularly since 1975, has now taken over the role of the smaller reactor.

To fulfil the statutory obligations of the Atomic Energy Authority including the operation of experimental reactors and provision of assistance to the CEGB and the Nuclear Power Company (NPC) in the construction and commissioning of the five AGR's, the Authority has eight establishments, each dealing with a specific group of activities. An outline of these provides an indication of their research and development capability, and, consequently, of that of Britain's nuclear energy industry.

Harwell is the largest research laboratory. About half its effort is directly concerned with reactor developments, while the remainder is about equally shared between work for the public sector and industry. This work is not directly related to nuclear but has been developed from it. Recently, Harwell's role has been expanded to provide support to the Department of Energy on general energy R & D matters. Notably, an Energy Technology Support Unit has been established which reviews alternative energy technologies, produces strategy papers and programmes of research in collaboration with the individual energy authorities for consideration by the Department and the Advisory Committee on Research and Development (ACORD), and is increasingly responsible for the management of R & D programmes in such fields as wave power. E.T.S.U. has also been playing a significant role in looking at prospects for industrial energy conservation. The Marine Technology Support Unit has served a similar role and now supports the Department of Energy's Off-shore Energy Technology Board, advising on and managing departmentally-funded programmes of research related to off-shore oil development.

At the Culham Laboratory, Abingdon, Oxfordshire, the AEA carries out its research in nuclear fusion, plasma physics and the technology associated with these. This laboratory also has strong links with outside industry, through contract research in such subjects as electro-technology and laser applications. Culham has been chosen as the site for JET (Joint European Torus), a major experimental machine designed to obtain and study a plasma in conditions close to those required in a fusion reactor. It will be financed and built as part of the Euratom Commission's fusion programme.

A number of facilities are located at Risley, Warrington. This establishment, known as the Risley Nuclear Power Development Establishment includes the Reactor Group, the Engineering Division, Central Technical Services and the Nuclear Power Development Laboratories. In addition to coordinating the programmes of the Dounreay Nuclear Power Development Establishment, the Windscale Nuclear Power Development Laboratories, the Springfields Nuclear Power Development Laboratories and those located at Risley itself, Risley maintains close links with British Nuclear Fuels and the Nuclear Power Company. Engineering Division is responsible for the design, procurement and inspection services for the Reactor Group while Technical Services similarly assist with studies and computer facilities.

Risley Nuclear Power Development Laboratories have an important role in nuclear developments since they are concerned with all stages of reactor component development. Their facilities include large-scale rigs for evaluating components, operating with sodium, water and high pressure gases. These are supplemented by laboratory facilities covering such areas as hydraulics, vibration, corrosion, metallurgy and pressure vessel technology.

Much of this work is closely related to safety, for which, however, the Authority has its own Safety and Reliability Directorate at Culcheth and also at Warrington. The Directorate advises the Authority on the formulation and application of safety and reliability policies. It coordinates general reactor safety research for the nuclear power programme as a whole and carries out safety research for individual reactor systems, particularly the fast reactor. It also undertakes studies on general safety matters on behalf of the Health and Safety Executive and, on payment, for industrial customers.

Each of the three experimental or prototype reactors has a research establishment associated with it. The Dounreay Experimental Reactor Development, for example, concentrates on fast reactor development, on the testing of fuel elements, fuel reprocessing and on development work on components required to operate in sodium.

At Windscale, there are facilities for the post-irradiation examination of gas-cooled and water cooled reactor fuels of all kinds while the Reactor Development Laboratory is also engaged on developing plutonium fuels for fast reactors. Development work on advanced gas-cooled reactors is carried out in conjunction with the Windscale AGR on the same site.

Atomic Energy Establishment Winfrith is mainly concerned with thermal and fast reactor systems, including such areas as reactor physics (notably the large zero-energy facility for fast reactor physics studies, ZEBRA), heat transfer and fluid dynamics and reactor safety and control. Other facilities are concerned with studies of reactor core performance and new reactor concepts. The prototype SGHWR is on this site.

Finally, at Springfields, Preston, the Reactor Fuel Element Laboratory deals with the development of fuels and fuel elements for thermal and fast reactors. Laboratory facilities cover the complete field from prototype manufacturing plant to special rigs for carrying out environmental tests on complete fuel assemblies.

Oil Development

Unlike the energy industries referred to so far, research and development in the oil industry is carried out by private companies which devote considerable resources to two main aspects of their work – developing new ways of exploring for oil and techniques for extracting and processing it, though, like the other fuel industries, the oil companies, at the marketing end of their activities, also undertake research into the better utilisation of their products, as motorists, to mention only one area, know from the periodic appearance of "new" petrols and lubricants, with claims of better mileage per gallon and other benefits.

Exploration for oil (and also for gas and coal) will always consist of two stages – a preliminary assessment by geologists and geophysicists, followed by trial drilling to discover if oil is really there and, if so, if it is present in exploitable amounts. Although these two stages must remain, research and development are helping to increase the accuracy of the preliminary assessments and to speed up trial drilling procedures. Much is also being done to improve the techniques for analysing the findings. Advances from other fields, such as new concepts in instrumentation, new analytical techniques and the application of computers are all making their contributions.

Research on oil production techniques includes work on methods for increasing the percentage of the oil-in-place which can be economically extracted from a field. One method tentatively put forward is the use of chemical explosives or even small nuclear devices to release additional deposits.

A related problem is the search for methods for the commercial extraction of oil from oil shales and tar sands. One method, which involves the mining of shale, followed by heating in a retort to extract the oil, generates large amounts of waste and uses vast quantities of water. An alternative proposal is to tunnel the heating retorts into the rock itself. Whatever techniques are eventually developed, it is clear that a great deal of expensive research will be necessary if oil shale and tar sands are to make a major contribution within the time scale envisaged in this book.

Another type of research and development, shared mainly between the oil companies and engineering contractors, is the new technology needed for exploiting offshore oil from ever greater depths and even more hostile environments than those so far encountered. The research covers all aspects of the operation, with particular emphasis on methods for extracting the oil, whether by steel or concrete platforms or by a form of sub-sea completion, when the well-head, instead of being on a static platform of steel or concrete, is actually located on the sea-bed.

At the same time, considerable attention is being paid to questions of safety and environmental protection.

Alternative Energy Sources

Although I do not expect the so-called alternative energy sources of solar, wind, tidal and geothermal energy to make a major impact on the world's energy situation in the foreseeable future, some reference to these longer term concepts should be included in a chapter on energy research.

Solar energy is undoubtedly the most promising of the four, and could it be harnessed, it has been calculated that only a fraction of 1 per cent of the solar energy received by the earth would meet the energy requirements of the whole world.

A considerable amount of research is in hand around the world, including the use of heliostatic mirrors (which move with the sun to reflect its energy on the same point at all times) onto a small boiler. Another technique, which is under investigation in the United States in particular, is a photovoltaic device which converts solar energy directly into electric current.

Less ambitious, but nevertheless useful work is being conducted on various forms of solar panels for heating domestic hot water supplies and swimming pool water.

Geothermal energy, in the form of superheated steam, if occurring near the surface, can be economically harnessed to drive turbines, and has been so used at Lardarello in Italy for 70 years, generating about 365 MW of electricity. There are similar systems in the United States, New Zealand and elsewhere. Hot water deposits are more abundant but technically more difficult to exploit though American research suggests that one possible technique would be to develop a suitable heat exchanger which would transfer the heat to more volatile liquid which could then drive a turbine.

Tidal power has also been around as a source of energy for a long time – the Severn barrage scheme, for example, was first mooted well over half a century ago. Schemes in operation include the Rance Barrage near St. Malo in Brittany, generating 240 MW, and the 400 MW installation across Kialaya Bay in the USSR, where larger schemes are being planned. Some very imaginative schemes have been proposed, including a series of vanes or "ducks", moored in the open sea, to convert wave energy into electric power. Putting forward this idea, S.H. Salter claims that a suitably shaped vane could remove over 90 per cent of wave energy and that a system of this type, stretching from Sule Skerry, 50 kilometres off the north Scottish coast, past the Butt of Lewis, Barra and Tiree to Islay, could meet all present UK electricity needs!

Some interesting suggestions have been put forward for new types of windmills, using materials and techniques developed in other areas, notably aeronautics. These suggestions are all speculative, and worth pursuing for their own sakes, but, facing the realities of the world's energy needs, one comes back to earth with the recognition that the major research effort must continue to be concentrated on the main existing fuels.

This is particularly true of research and development programmes which relate to the world's coal industries and the world's governments will do well to make sure that the necessary resources are made available from now on, if future energy requirements are to be met in full.

9 Energy and the Environment

As if the mounting problems of locating new sources of energy, extracting them and putting them to best use were not sufficient, those whose task it is to seek to give the world the energy it needs are becoming increasingly accountable for the effects their actions may have on the environment. Yet, though it adds a new dimension to the problems, it is a fact of life which we all have to accept since, in our several ways, we are all equally anxious to preserve the environment – for our own sakes as much as for those who will follow us.

Problems of the environment are much greater than ever before, not only because people with a particular interest in the quality of life are more numerous and more vociferous and much better organised than previously but also because the term "environment" is defined in a much broader sense. In addition to objections raised on visual grounds, present-day environmentalists (and who shall blame them?) will oppose plans which are likely to pollute the earth, the air and the sea, by liquids, solids or gases while, in recent years, there has been increasing attention to what has come to be known as the "fourth pollutant" – noise.

Almost every aspect of the production, processing, distribution and usage of fuel can potentially harm the environment, as defined above, in one way or another.

Visually, the environment is most obviously affected by such installations as coal mines and open-cast sites, oil refineries, power stations, electricity transmission lines, natural gas terminals and storage facilities.

Air pollution can arise from the use of fuel, in the form of discharges of such pollutants as carbon monoxide, carbon dioxide, oxides of nitrogen, sulphur oxides and lead. Dust and grit may also be emitted to the atmosphere from factory chimneys; they stay in the air for a shorter time but their effects are nonetheless unpleasant.

Oil spillage from tankers is one of the more dramatic forms of water pollution and to this must now be added the risk of oil leaks from offshore installations and undersea pipelines. The discharge of cooling water from power stations is a possible cause of thermal pollution although it is now questioned whether any particular harm results from this.

Most of these environmental hazards have been with us for a long time and this also applies to the consumption of fuel in the transport field, leading to unpleasant fumes from diesel engines and potentially harmful discharges from petrol engines, though the degree of risk is a matter of keen debate between the experts.

Aircraft and road vehicles also add to the problem of environmental noise, though whether this can be laid fairly at the door of the fuel must be open to question, the mechanical design of the machinery being at least partly to blame. On the other hand, the energy industries can be directly responsible for some noise, from electrical transformer noise to the noise of gas turbines at natural gas pipeline booster stations.

A form of environmental pollution which has assumed great importance in recent years and which may have a very big influence on the future development of the world's energy resources, particularly in the longer term, is the risk of radiation from nuclear power stations, nuclear fuel re-processing plants and other installations. President Carter's policy pronouncement in April 1977 about the security risk of plutonium resulting from the operation of re-processing plants and fast breeder reactors has brought a new topic to the nuclear debate.

Yet if the development of nuclear power seems to be arousing the most serious and widespread objections, the other forms of pollution I have mentioned also have their determined opponents, particularly in the localities where the immediate effects will be felt, such as the area where a new coal mine is to be exploited or a natural gas compressor station erected.

From this brief run-down of pollution worries, it is clear that everyone concerned in meeting the world's energy needs has serious environmental problems to solve.

The Dilemma

As previous chapters have shown, the world must have energy if living standards are to remain stable, let alone rise. This will mean, as I hope I have clearly shown, that massive exploitation of all the main forms of energy will be required and this will be impossible, certainly on the scale required, without some adverse effects on the environment.

We all need the energy, even the most dedicated of environmentalists, whose worthy schemes for improving the environment may even demand additional energy. Nor must it be assumed that those of us who are involved in producing the energy are unmoved by the environmental effects of our efforts. This particularly applies to coal mining when those who produce the fuel have to live in the area where the results of their work are often only too apparent.

Here, then, is the dilemma. Energy we must have and yet to produce it inevitably affects the environment unfavourably. Of course, there are many steps which can be taken to reduce the harmful effects of energy production and much of this chapter will be taken up with discussing these measures. As a general rule, however, the greater the care that is taken to control pollution, the greater will be the cost of obtaining the energy.

Taken to extremes, it might be possible to extract coal, oil and natural gas only in places so remote from centres of population that no one would notice the environmental effects, or to construct all power stations underground, but the cost would be enormous. At the other extreme, there is the example of the Industrial Revolution when energy (primarily coal) was exploited wherever it could be extracted and used most economically, with environmental results which were serious at the time and still linger on in some areas.

Somewhere in between these two extremes we shall have to make a compromise – between the world's urgent need for energy and the equally urgent need to protect the human environment.

I believe we have been moving towards such a compromise for some time past, thanks, on the one hand, to the actions of government, local authorities, interested organisations and pressure groups and, on the other, to the work and concern of the fuel industries. The events of the past four years since the 1973 crisis, by bringing the energy problem into sharper focus, have directed a spotlight on the environmental aspects of future energy policies, providing an opportunity to consider what is being done and what more could be done in the future.

It is a very big problem which can only be touched upon here, where I propose to deal with the major areas of environmental concern and indicate how the energy industries are taking steps to minimise the effects of their actions.

The Physical Environment

I begin with the effects of energy production on the physical environment – an aspect of the general problem which probably arouses as much opposition as any other because it is immediately noticed by the eye. It is also an area in which we in the coal industry once had a traditional role as a despoiler of the landscape but, in recent years, have transformed ourselves into what I believe is now one of the most amenity conscious of all modern industries.

Today, we are firmly committed to the protection of the environment at all stages of the mining, processing and transport of our products. Concern for the environment covers both remedial action at existing collieries, to alleviate the worst effects of earlier neglect, and care in the opening up of new reserves. Our policies cover open-cast sites as well as deep mining, but it is important to stress at the outset that our firm commitment to protecting the environment has to be considered in relation to our no less important duty to produce the coal the nation needs, in sufficient quantities and at acceptable prices.

At existing mines, the main environmental considerations relate to new surface structures, subsidence and the disposal of the large quantities of solid waste which are inseparable from most mining operations. We at the NCB maintain close and continual contacts with local authorities when planning surface developments as well as submitting to the more formal control of the Town and Country Planning Acts. The increases in coal output under our long-term plan will require considerable additional facilities of all kinds, from coal preparation plants to marshalling yards, but all such structures will be designed to the best standards of modern industrial architecture.

This is a policy we have been following now for a number of years, as anyone who has seen any of our recent colliery buildings must surely agree. In all our new construction work, we set ourselves high standards.

Subsidence above new workings is no longer the problem it once was, due to improved mining techniques and to the fact that, in general, we are now working coal at greater depths. Much of the success must be attributed to the careful research which has been devoted to mining processes which has directly or indirectly contributed to the reduction and the control of subsidence.

As an example of what can be achieved in this direction, there is the case of Peterlee New Town, County Durham, which was actually being created on the surface while 8,500,000 tonnes of coal were being mined from beneath the designated area. No less than 3,500,000 tonnes of this coal were taken from beneath built-up areas, both residential and industrial.

This was achieved with minimum disturbance by close cooperation between the NCB and the Development Corporation.

Colliery tips were formerly the most damaging to the environment of all the effects of coalmining and the disposal of waste remains a continuing problem. A large amount of solid waste is an inevitable accompaniment of most mining operations and, as has been all too painfully obvious at many points during the earlier history of coalmining, colliery tips can be a source of nuisance or even considerable danger through instability, fire, erosion, effluent pollution, blockage or flooding.

It is obviously beyond the capability of the NCB to dispose entirely of the 2,000 million tonnes of waste deposited in the past, but it deals effectively with the old tips in its ownership. Some of these tips are "re-washed" to recover the coal they may still contain, and the residue so obtained is deposited in new and more acceptable places.

About 4,000 acres of previously derelict land have been transferred from the NCB to local authorities who, by using government grants that are not available to the NCB can rehabilitate old tips and areas of subsidence. A notable example is the Strathclyde Regional Park (Scotland) where a remarkable new recreation area has been carved out of derelict land over a former coalfield, the last colliery of which, Bothwellhaugh, was closed in 1958. A major feature of the Park is a huge man-made loch suitable for all forms of watersport, including competitive rowing events at full Olympic status.

Well established and scientific procedures are now strictly enforced by the NCB to prevent the risks which formerly attended tipping and, in some cases, tipping is actually used to improve the amenities of an area.

For example, at New Stubbin Colliery on the Yorkshire coalfield, a sloping valley site of comparatively poor land was levelled by controlled tipping and the surface laid out as playing fields for a local West Riding County Council school. I could quote other examples to show what can be done to lessen the harm to the environment caused by modern mining methods. I could also refer to the use which is often made of colliery waste in civil engineering and building, such as its use as a "fill" for roadworks or other projects, or for use as a component of lightweight building aggregates.

Opencast mining methods, being of more recent introduction, at least on any scale, do not have as grim a history of despoilation as deep mining. Even during the Second World War and just after, when opencast methods were employed almost as a means of national survival, care was always taken to cause only the minimum amount of disturbance to the life style of local inhabitants and their amenities.

Since then, a great deal has been learned about the science and practice of land re-instatement, by the NCB and others concerned with this problem, and today the greatest possible care is taken to restore the land to use again as soon as possible after the coal has been extracted. In many cases the land is left in a better state than before operations were begun.

This is especially the case when opencast mining takes place alongside former deep-mining operations, when former waste tips can be lifted and reformed in conjunction with the overburden from the opencast workings. In all opencast operations, the full planning procedures have to be followed and the closest co-operation maintained with local authorities and other interested organisations, especially when planning the reclamation of an area.

"Plan for Coal" looks to an expansion of opencast coal output, as we have seen, from the present annual total of around 10 million tonnes to 15 million tonnes a year and this will inevitably mean the temporary occupation of a considerable area of land while the coal is extracted. I can well understand the objections which local residents put forward and I have more than once had personal experience of the strength of feeling which can be aroused!

Exploratory drilling is taking place in many areas but no coal production takes place until the full procedure for permission has been completed and the Minister only gives the go-ahead when he is convinced of the justification of the case for mining.

Once permission is granted, we go to considerable lengths to carry out the work with as little disturbance and loss of amenity as possible.

I can illustrate this by reference to the 2000-acre Butterwell site near Morpeth, Tyne and Wear, where permission to start work was only given after the holding of three very detailed public enquiries.

Some hard things have been said about this particular development and I fully appreciate the fears of those who are concerned about the local environment. I hope, and believe, however, that people living in the area will come to see that their worries were largely unfounded as they see the care we take to ensure minimum possible inconvenience during mining and that subsequent re-instatement practices are of a high order.

Included in the steps we shall take will be the construction of baffle mounds which will blend with the surrounding landscape and reduce both the visual effects and the noise from site working. After construction, these mounds will be sown with grass to merge them with the Butterwell landscape.

We shall be using a cyclic system of working, in which the spoil from one strip is used to fill and re-instate the previously worked-out strip. The working area exposed at any one time is minimised, ensuring that the land is put back to its former use in the shortest time possible. Before the Butterwell site is even half worked out, the first strips will have been fully restored, with animals grazing and crops growing once more.

There will be constant monitoring of the restoration programme, under the control of the NCB and the County Council, with regular consultation on the siting of trees, hedgerows and other topographical features. We regard ourselves as responsible environmentalists, aiming to leave opencast sites in at least as good condition as we find them, and – more often than not – better.

At the same time, we are not so naive that we believe we can ever develop opencast sites without some disturbance for the people in the locality. Yet we believe we can be accepted as good neighbours who are doing a job of economic importance. In the case of Butterwell, this means 12 million tonnes of high-quality and economically produced coal over the next ten years.

Coalfields of the Future

Environmental problems which arise from opencast mining are temporary – though I accept that a ten-year project as Butterwell is may not seem to be temporary to local inhabitants – and the effects of construction work and the disposal of waste in existing mining areas are somewhat mitigated by the fact that there are other mining works there already which, to a certain extent, softens the impact of the latest work. What of the entirely new coalfields which will be opened up – the Selbys, the North East Leicestershires and the others which will be needed as the years go by?

Here, too, there has been, and will continue to be understandable opposition and only after prolonged enquiry and investigation have we been given permission to work the Selby coal and this will also be the pattern when we decide to seek permission to develop the Leicestershire deposits.

At Selby there will soon be an opportunity to see how a modern coalfield can be developed with minimum disturbance to the environment and I am confident that we can exploit these rich reserves in such a way that people living on other potential coalfields will find their own fears allayed.

Let us look at the environmental aspects of the Selby project.

Although the coalfield covers over 100 sq. miles, the surface buildings will occupy only five sites – four of these comprising the shafts for ventilation and for raising and lowering men and equipment and the fifth the drift mouth where coal will be brought to the surface and where loading plant will be sited.

As I explained in an earlier chapter, the drift mouth will be located at the site of a disused railway marshalling yard at Gascoigne Wood, on the main railway line west of the town of Selby, and so will be an addition to an existing industrial installation and not a brand new feature. Gascoigne Wood will therefore revert to its original use, though on a considerably larger scale than before. A detailed survey of the situation at an early planning stage had confirmed that Gascoigne Wood was the most acceptable site, environmentally, and, as it happened, operationally, from the several alternatives considered.

All the stone from the drift and tunnels will be formed into landscaping screens which will virtually surround the site. Suitable plantings will be made on and around the embankments to help merge the area into the rural atmosphere. The end result may well be a more environmentally acceptable locality than the present disused marshalling yard!

As all the coal will leave this depot by rail, and there will be little additional burden on the local road network, once the major initial construction work is over, we shall be meeting one of the major areas of concern revealed at the Public Enquiry. Another series of objections related to possible harmful effects on the landscape of this predominantly rural area, and we believe we shall be satisfying objections on this score by careful design of the four mine shaft installations.

For a start, we have agreed that the height of the tallest winding tower at each site shall be less than 97 ft. NCB engineers have designed a type of winding engine and rope system which will keep the height of this equipment as low as possible, and certainly within the limit laid down. Other buildings on each site – they will range from 20 to 40 acres in extent – will include offices, baths, a medical centre, canteen, workshops, stores, stockyard, car and bus park and electricity substation, but none of these will be more than 70 ft. high.

It is our belief that after completion of construction, there is no reason why the sites should cause any more intrusion into the surrounding landscape than a moderate sized light engineering undertaking. Much of the area even of the smaller sites will be taken up by landscaping, designed to merge the whole installation into the surroundings.

Environmental problems encountered at Selby relating to the removal of the coal and the design of buildings can be solved in a similar manner at other future coalfields – in North East Leicestershire and elsewhere–with the added advantage that, with later fields, we shall have the benefit of the experience gained at Selby.

Every new coalfield throws up its own special problems and two which were of special significance at Selby were the risks of subsidence and flooding.

Subsidence, which, as I have already said, presents fewer problems with modern mining techniques, is being strictly controlled everywhere on the field while pillars of supporting coal will be left under Selby itself, so as to ensure that no harm comes to the beautiful Abbey.

Fortunately, the excellence of the reserves below the ground make them ideal for what we describe as "in seam" retreat mining and a pillar/panel system of extraction which combines maximum production efficiencies with a controlled degree of subsidence. The pillar/panel ratio has been calculated to give the required stability to the headings and at the same time control subsidence.

We were similarly able to deal with the criticisms of those who feared that our mining operations might lead to flooding in an area of flat, low-lying land, crossed by two rivers. Some of the first contacts we made in the area were with the Yorkshire Water Authority, with whom we have agreed on the necessary precautions to be taken to eliminate flooding risks.

Gas

We in the coal industry are not the only energy industry to cause environmental problems nor are we the only people to be facing up to our responsibilities in this regard. The gas industry, too, have to satisfy the environmentalists and, like ourselves, pride themselves on their efforts. In the past, like us, the gas industry had come in for a deal of criticism because of their effects on the environment, mainly the appearance of the old-fashioned coal gas works and gas holders.

More recently, the newer reforming stations, being oil based, took on the appearance of oil refineries or petrochemical plans rather than the traditional "gas works" of former times. These latter were increasingly dismantled (curiously enough, to cries of dismay from some environmentalists who regret the passing of industrial archaeological relics which should be preserved!). With the arrival of North Sea gas different environmental effects have arisen, in the form of pipelines, reception terminals and gas booster stations at intervals along high pressure transmission lines as they cross the countryside.

British Gas is no less concerned at the effects of its activities on the environment than any other energy industry. Its concern begins with the laying of pipelines when great care is taken to reinstate the land following the not inconsiderable but temporary disturbance caused by the laying of a major pipeline. There is pretty fair agreement that the pipelining programme of British Gas has been one of the major environmental successes of recent civil engineering.

To help ensure this, British Gas set up an Environmental Planning Department which plays an important part in the selection of sites for the construction of major gas plant, such as shore-based, gas storage installations, the gas booster stations and others.

Informal discussions are held with officers of the local planning authorities before a specific site is selected and only when a possible location has been chosen which raises the minimum objection from the planners and which is technically suitable, is the formal application for outline planning permission submitted.

Detailed discussions then take place with a wide range of interests, with particular reference to local residents. Bodies involved include both statutory organisations, such as the Countryside Commission and the Nature Conservancy Council, which have legally to be consulted, and others, like the Council for the Protection of Rural England (Scotland and Wales), the Royal Society for the Protection of Birds, the Wild Life Trust and the National Trust, which are consulted on an informal but nevertheless important basis. Others consulted include the National Farmers' Union, the Country Landowners' Association and, where appropriate, the Forestry Commission.

Following all these discussions, outline planning permission is obtained, to be followed in turn by detailed planning consents, covering such matters as overall site layout, the means of access, the appearance of any buildings, external colour schemes, artificial lighting, proposed landscaping, etc.

Special attention is paid to noise problems, and, in particular, the noise from booster stations which can cover the whole spectrum, from high frequency levels to the lower frequencies which are particularly difficult to control. British Gas has pioneered a new approach to the problem by enclosing the compressors in an acoustic enclosure or "cab".

I realise that I have only given an outline of the environmental activities of the gas industry – sufficient, however, to underline the point I have already made in connection with the environmental problems of coal, namely, that a balance must – and can – be struck between the apparently irreconcilable aims of those who seek to protect the environment and those whose task it is to ensure the nation's energy supplies.

Electricity Generation

It is symptomatic of the growing concern of governments at the effects of the energy industries on the environment that when the Central Electricity Generating Board was established in 1958 it was given the statutory duty of taking into account, when carrying out their duties, any effects its activities might have on the natural beauty of the countryside, including the flora and fauna as well as buildings and objects of special interest.

Although coal mining has to be carried out wherever the coal exists in the ground, which may be where there are surroundings of great natural and other interest, the siting of a power station, especially in a country of our size; is not tied so strictly to a particular locality. However, the technical requirements of a modern electricity generating station are such that there is less flexibility than formerly in selecting the optimum site.

For example, coal-fired stations must be reasonably near to an economic source of coal in large quantities. This either means a site over a coalfield or with very good access by rail. Facilities are also required for the large scale disposal of the pulverised fuel ash, either for transformation into a commercial product or as landfill.

Oil-fired stations have to be sited where there is access to a deep water anchorage where tankers can discharge, or within pipeline range of a refinery.

Much more stringent requirements apply to the site of nuclear power stations, but I shall return to this point later in the chapter.

All types of electricity generating stations other than gas turbine installations must obviously be sufficiently near a large and unfailing supply of water for condenser cooling. A power station occupies a large area of ground and this must be reasonably flat and above flood level and yet low enough to avoid heavy cooling water pumping costs. The ground must be capable of supporting the heavy structures and there must be good roads for the heavy flow of construction traffic, which will continue for a number of years.

When all these parameters are considered, it is clear that suitable sites for new power stations may at times be proposed in areas prized for their amenity value.

In an attempt to overcome objections, projects are planned from the outset in consultation with all planning and amenity interests, in order to cause the least possible disturbance to the environment. The Board makes its decision on a choice of site for a proposed power station development only after exhaustive investigations and consultations with a wide range of statutory and other bodies. The Board's final choice represents their judgement of the site which offers the best compromise of amenity, technical and economic requirements.

Once the site is agreed, environmental considerations continue to dominate the design of the station, with close attention to site layout, the orientation and shapes of the buildings and the use of texture and colour in the construction of the major buildings and structures. Natural draught cooling towers pose a particular problem and although studies of grouping and massing can do much to mitigate the visual impact, a drastic reduction in the visual impact of cooling tower installations has been made possible by the Board's development of the assisted draught cooling tower, one of which is capable of performing the duty of three or four natural draught towers.

Environmental debates also take place over the design and siting of overhead transmission lines which can detract from the visual appeal of the countryside over which they run. Much is done to reduce the adverse effects, through comprehensive consultations with the bodies concerned.

Great thought is given to the route followed by the lines with the aim of fitting them as unobtrusively as possible into the landscape. Techniques employed include the avoidance of skylines, the making of detours to avoid spoiling fine views, taking advantage of broken country to form a background and using the natural screening provided by woodlands and hedges.

Many of the measures, such as following a route other than the direct one, for environmental reasons, add to the cost of the project, and the nation as a whole has to decide, once again, on the relative values of energy and amenity.

It might be thought that the whole problem could be avoided by laying all power lines underground but both the cost (in the case of 400 kV underground cable it is 15 times the cost of 400 kV overhead line) and very serious technical problems make this impossible. For amenity reasons a very limited amount of 400 kV underground cable has been installed in seven locations. In many cases 132 kV line has been put entirely underground in order to allow a 400 kV overhead line to be built with the minimum of environmental impact.

Like the coal industry, the CEGB is continually seeking methods for disposing of its waste products – in this case, the ash from coal-fired boilers – without detriment to the environment. In addition the Generating Board takes very great care to minimise the effects of chimney emissions through the use of very high efficiency electro precipitators, their pioneering of single tall multi-flue stacks and the continuing programme of monitoring and research and development undertaken by the Board in this field.

Pulverised fuel ash (PFA) has long been used as a landfill material and when disused quarries or gravel pits are filled and reinstated as agricultural or other land, there is a net gain to the environment.

More recently, PFA has been put to commercial use, in the form of motorway embankments, the manufacture of building blocks and lightweight aggregates, as well as a number of more specialised uses in concrete, grouting, soil stabilisation products and as industrial fillers. About 60 per cent of PFA is now used in these and similar applications, paralleling the efforts of the NCB to dispose of colliery waste material, with benefit both to the environment and to the national economy.

Air Pollution

I would not like it to be thought that I am attempting to give a definitive account of the environmental effects of energy in this chapter. It is only my purpose to outline some of the considerations which will arise as we develop the increased energy resources we shall need. Air pollution, for example, is far too big a question to be covered fully in one section of one chapter.

However, since the combustion of all fossil fuels can produce by-products which may have harmful effects, if present in the air above certain levels of concentration, I make some reference to this subject here. In fact, fuel combustion is generally reckoned to be the main source of man-made air pollution throughout the world.

In the United Kingdom, action against smoke has been a major area of Government activity, especially since the Clean Air Act of 1956. The initial target was the reduction of industrial smoke, caused in the main by the incomplete combustion of coal in inefficient boiler plant and furnaces.

Following the widely admitted success of the measures to control industrial smoke emissions, attention was then directed to the domestic field. Control of smoke from domestic premises has been left to local authorities, with financial support from the central Government.

Research and development and the manufacture of various types of smokeless fuels by the NCB has played a big part in the progress of domestic smoke control. Another factor has been the development of solid fuel appliances such as the "smoke-eater" type which burn smoke particles to prevent them escaping to the atmosphere.

Smoke emissions in the United Kingdom have fallen by some 60 per cent since the introduction of the Clean Air Act. The once notorious London "smog" has disappeared, as has the pall of black smoke over the industrial Midlands which gave it the now undeserved name of the Black Country.

Sulphur dioxide was another component of London "smogs". That combination of smoke and sulphur dioxide was definitely harmful to health but there is no certainty that sulphur dioxide in the concentrations found in Britain these days is injurious. Since reaching a peak in 1961 sulphur dioxide emissions have tended gradually to fall.

United Kingdom policy has been to avoid high sulphur dioxide concentrations at ground level by using tall chimneys which ensure rapid dilution of emissions. Sulphur dioxide remains only a short time in the air so there is little fear of serious cumulative effects. Indeed it has been claimed that airborne sulphur, formed by transformation of the sulphur dioxide in the air, can be useful in areas deficient in natural sulphur. But recently there have been suggestions that under certain meterorological conditions sulphur compounds can be airborne for considerable distances and that other countries may suffer deleterious effects from the sulphur coming from Britain's chimneys. This is, however, a far from straightforward problem since the statistical evidence on which these claims have been based does not always stand up to close examination, and the quantities of sulphur compounds concerned and their origin are open to dispute. What is clear, however, is that Britain is not the only country involved in this sort of trans-frontier traffic.

Methods of securing substantial cuts in emissions from power stations and other large industrial plants are under development in case they become necessary. We at the NCB for example are developing the technology of removing emissions from flue gases, while fluidised combustion, as explained in the research chapter, also helps to reduce sulphur emissions.

For oil, tall chimneys can, as with solid fuel, be used to ensure rapid dilution of sulphur emissions. Alternatively the sulphur dioxide can be controlled by removing a high proportion of the sulphur from distillate products, such as kerosene and gas oil, to prevent high sulphur dioxide concentrations at ground level from space heating, both domestic and commercial.

Carbon dioxide is the final product of the combustion of all conventional fuels, and, since it is already abundant in nature and is in any case essential for the maintenance of life, it only becomes a pollutant if present in such large concentrations in the atmosphere that it interferes with the natural processes. Present levels are far below those which have adverse effects on health, but there have been suggestions that the increasing concentrations of carbon dioxide in the atmosphere could lead to a general warming of the climate. The World Meteorological Organisation monitors the carbon dioxide content of the atmosphere on a global scale to keep a watch on possible effects on weather and climate.

Carbon monoxide, resulting from the incomplete, and therefore inefficient combustion of conventional fuels, is in an entirely different category. It is toxic and, as is well known can be fatal when present in confined conditions.

Internal combustion engines running on a "rich" mixture, can cause concentrations of carbon monoxide at ground level, but, except in very congested conditions, health risks from this cause are not considered serious.

Considerable concern has been expressed in recent years about the risks of lead pollution from vehicle exhausts. Fossil fuels themselves contain very little lead, but lead, in the form of tetra-ethyl or tetramethyl, is added to petrol at the refining stage to improve its octane rating. The quantity of lead compounds emitted in vehicle exhausts is very small but lead is highly toxic, persists for a considerable time in the atmosphere and can accumulate in humans, animals and vegetation.

Lead pollution is a highly emotive subject and there is a strong movement against leaded petrol, with demands for its abolition, or at least restriction, by law.

This is only one aspect of the wider problem. Air pollution in all its forms has been with us for a long time, and it can be assumed that action to lessen its effects will continue, both at government and industry levels. The increasing demand for energy may increase the risk of air pollution but I suspect that this will be more than counterbalanced by action to reduce it by legislation which in turn will lead to improved techniques.

Water Pollution

Water pollution has been the subject of legislation in the United Kingdom ever since 1876 when the Rivers Pollution Act first brought the law into play. The most recent enactment, the Pollution Act 1974, extends controls to cover virtually all discharges to inland and coastal waters. The energy industries are far from being the worst offenders and the coal industry makes very little contribution to water pollution.

There are liquid wastes from mines and coal washeries but we have developed satisfactory methods for dealing with these and I am not aware of any serious complaints on environmental grounds.

Less happy is the state of the oil industry which can cause considerable pollution hazards, especially accidental oil spills from tankers, of which there have been some spectacular examples. There are also the largely unpredictable risks of large-scale leakages from North Sea and other offshore oilfields, which were highlighted in the early part of 1977 by the blow-out in Norway's Ekofisk field.

A risk of accidental spillages of oil at sea will have to be accepted as long as vast tonnages of oil are shipped around the world in tankers. Steps can be taken to minimise the risk of collisions involving tankers by improved navigation arrangements (the one-way system in the Straits of Dover is one example) whilst developing methods for clearing up the mess should an oil escape occur.

Techniques that have been developed for dealing with oil spillages include the "sand-sink" method under which the oil slick is sprayed with a mixture of sand and water. The oil adheres to the sand and sinks to the sea-bed. Other methods are based on the use of low toxicity detergents which divide the oil into small droplets which are naturally dispersed by time, currents and wind. Yet another approach is to contain oil spills by an inflatable boom which collects the oil in a confined space for easy disposal.

Much valuable work along these lines has been carried out by various organisations including Warren Spring Laboratory at Stevenage. The methods are there, the question now being to ensure that the necessary materials and equipment are available and the organisation is in existence to take action quickly when a major emergency occurs. The Ekofisk incident, when a valve failure resulted in vast quantities of oil being discharged onto the surrounding sea, will have made the offshore operators and governments concerned even more conscious of the need to ensure that the procedures for cooperation in emergency measures will prove adequate when the need arises.

These risks are a direct result of the need to develop new sources of energy and, granted that we need North Sea oil, as we certainly do, we must accept the risks.

Oil refineries, too, have their effluent problems, through the discharge of oil in certain effluents, such as process steam which, having been mixed with oil vapours during refinery operations, produces a mixture of oil and water on condensation. Drainage waters from paved areas of a refinery can also prove difficult to dispose of without causing an effluent problem.

Effluent treatment techniques can adequately deal with this type of problem – but at a cost. It is reckoned, for example, that up to 10 per cent of the total capital cost of a modern refinery goes on the control of water and air pollution. The cost of pollution control, in all its forms, is added to the price of energy, thus increasing still further the cost of future supplies, already likely to rise because of dwindling supplies.

Noise is another environmental hazard which is arousing growing objections. Mining no longer causes serious noise problems on the surface (noise below ground is a different matter) although plants engaged in various forms of coal processing and production can cause the same problems to people living or working nearby as other factories where heavy machinery is used.

Oil refineries can have noise problems, especially when commissioning new plant, and power stations can emit some noise, but few refineries or power stations are near enough to centres of population to raise many objections.

Big strides have been made in recent years in the development of techniques and equipment for the control of noise. Here, too, the costs of control are high and, once again, the public has to weigh the additional cost of a better environment against the world's need for energy.

Nuclear Power and the Environment

The environmental implications of nuclear power have been and will continue to be debated extensively. Even now, when nuclear power represents only a small (though growing) part of the world's total energy production, there is considerable, and growing, concern on the subject.

Great care has undoubtedly been taken to ensure the most careful control over all aspects of nuclear power generation. Any human activity carries risk and nuclear power generation is no exception. Research on safety and environmental protection are designed to reduce any risks to a minimum and to establish the most efficient ways of dealing with waste and the decommissioning of obsolete reactors.

During normal operations, there is a minute discharge of radioactivity through the stacks of nuclear power stations, at about 200 ft. above ground level, and also a very small discharge in liquid effluents. Careful monitoring indicates that the amounts of radioactivity are far below permissible, safe minimum levels and, indeed, it has been claimed that discharges from nuclear reactors, in normal operation, are potentially less harmful than the pollutants discharged from conventional power stations.

Dr. Edward Teller, Professor Emeritus, University of California, speaking in London on October 21st 1976 said: "A very great number of careful safety measures and safety investigations have been carried out, and as a consequence in the United States we have today almost sixty big nuclear reactors, many of which have been working for years. Not a single one of these has damaged the health of any individual as far as we know. This is a safety record unparalleled by any other method for producing energy."

In the UK, too, extreme precautions are taken to ensure that reactor accidents are improbable and that even if a serious malfunction should occur, fission products or other radioactivity would be contained. The Nuclear Installations Inspectorate, a branch of the Health and Safety Executive, have the responsibility of ensuring that all reactors for the Generating Boards are designed, constructed and operated to the highest possible standards of safety. To this end they are empowered to withhold a licence to operate a reactor if they are not fully satisfied on all points of safety. Similar arrangements apply for British Nuclear Fuels Ltd's fuel plants.

Some of the most urgent environmental problems surround the reprocessing of irradiated fuel elements and the consequent recovery of residual uranium and plutonium which is an important part of the nuclear fuel cycle. The UK began this activity at the Windscale plant some 25 years ago, as part of the defence programme and has been reprocessing nuclear fuel since the commissioning of Calder Hall in 1956.

Guiding principles have been developed for the safe disposal of wastes in the UK. Gaseous wastes, which arise when fuel is dissolved and from the ventilation of cells and vessels, are treated by high efficiency filtration, inertia demisting, electrostatic precipitation or scrubbing or absorption processes. All these are well-proven processes, the choice depending on the nature of the contaminant to be controlled.

There are two main categories of low active solid waste. The first consists of such materials as clothing, paper and equipment which should be free of activity, but as it originates from the active area of the plant it is buried in approved areas. The second category comprises paper, gloves and items of equipment which have either had direct contact with radioactive materials or are known to be contaminated. Some of this waste, containing low levels of alpha contamination, is treated and packaged for disposal in the deep ocean, under the auspices of the Nuclear Energy Agency of the OECD.

Low active solid waste containing larger quantities of plutonium is stored at present while awaiting the introduction of suitable decontamination techniques. Various methods are under investigation but in the meantime the waste is safely buried.

Highly active solid wastes which comprise mainly fuel element cladding which has either been removed from the fuel mechanically or remains at the end of the acid leaching process, are at present stored under water to avoid a fire hazard. This procedure is fully satisfactory but further methods for future treatment are currently under investigation.

Techniques have also been developed for the disposal of liquid wastes, employing methods considered appropriate for the three classes of waste – low, medium and highly active wastes.

Low active waste is disposed of at sea under authorised procedures but a variety of methods is used for medium wastes, depending on their nature. Some are concentrated by evaporation, and the concentrate stored for a number of years to allow the relatively short-lived radioactivity to decay, when it can be dealt with as low active waste. Other medium active wastes are treated by a precipitation process which leaves a liquid, which can be treated as low active waste and a sludge which is stored in tanks for further treatment, packaging and disposal later.

Highly active liquid waste is concentrated and stored in high-integrity tanks within thick-walled concrete cells which are themselves clad in stainless steel. Internal cooling coils remove heat caused by the decay of the fission product and an external cooling jacket provides a further means for controlling the heat generated and also provides additional cladding for the tanks.

I have given this summary of waste disposal practices because it is around these that the main objections are put forward in many parts of the world, and have resulted in slowing down nuclear programmes in the USA and West Germany, where reactions have been particularly strong. The objectors base their case, not on the safety record of the nuclear industry to date, which can hardly be faulted, but on what may happen if the programme of nuclear power station construction is expanded and much larger quantities of radioactive materials have to be handled, transported and stored than at present. The implications of fast breeder technology are particularly feared and have been publicly voiced by President Carter although his particular concern is to prevent proliferation of material which is potentially useful as an offensive weapon.

For its part, the nuclear industry contends that technology is not standing still, including the technology of dealing with radioactive wastes. They would claim that the existing methods of waste disposal are perfectly satisfactory but that they are not necessarily optimal and that even greater safety would be achievable by the time new generations of nuclear reactors are ready to "go critical", say, a time scale of 10 to 15 years.

All this shows, of course, is that there is a considerable division of opinion on the subject which governments and the nuclear industry still have to resolve.

Striking a Balance

Concern about the effects of nuclear power programmes on the environment illustrate the general problem which faces the world – how can we strike a balance between the urgent need for energy and the effects the various proposals to solve the energy problem will have on the environment?

In Western Europe as a whole, a considerable amount of anti-pollution legislation has been passed and more is planned, but no common pattern has yet emerged which must presumably await agreement within the EEC. In the different countries, there are wide variations between the steps taken, depending on the pressure of public opinion, the rate at which the control measures are introduced and the effectiveness of enforcement.

However it is already clear that the measures so far taken, modest though they are, in some cases will add to the costs of energy and extend the time required to obtain planning clearance for new energy projects. The latter point is of particular relevance to coal mining, both opencast and deep mining, and to the construction of new power stations, conventional and especially nuclear. The delays may well hold down coal production, particularly opencast, below the levels which would otherwise have been reached by the various target dates.

I have restricted myself mainly to our domestic environmental problems in this chapter, chiefly because of the difficulty of dealing fully with so vast a problem on an international basis but also because the problems we face are encountered elsewhere in the world.

That the effects of environmental problems on future energy supplies have major international repercussions can be illustrated by reference to a Report of the National Petroleum Council in the United States in 1972. This stated that delays in the authorisations for the Alaskan pipeline system had deprived the nation of at least 100 million tonnes of crude oil a year and about 72 million tonnes a year of oil equivalent of natural gas.

New anti-pollution legislation, the Report stated, would mean that over 40 per cent of the estimated United States coal reserves lying east of the Mississippi River, with their sulphur content of over 3 per cent, would be unusable as a boiler fuel under the regulations in force in most places, until an economic means of using high-sulphur coal was developed.

Though these are domestic United States issues, the point I am making is that any decisions which reduce US home-based energy sources will increase her need for imports, making it still more difficult for the rest of the world to obtain its requirements from the dwindling reserves available for export.

As in all other aspects of the energy problem, the issues are international. We are all in this together and must individually, and collectively where possible, devise agreed means for developing the energy resources we need whilst at the same time paying due regard to environmental considerations.

10 Policies and Prospects

The time has come to sum up and conclude.

During the course of this book I have described the world energy situation as it has developed since the last war until the present time. I have emphasised the significance of the action taken by the OPEC countries in October 1973 and subsequently in massively increasing the price of oil. In the period of some four years that has elapsed since then it has become clear that this was no passing phenomenon, for, in spite of the recent world economic recession, the price of oil has stayed firm.

What is even more important is the likely trend in the energy situation from now till the end of the century – a matter of just over twenty years. The OPEC countries in 1973 increased the price of the most important fuel moving in international trade for political reasons and through the strength of their monopoly position – no real shortage has developed. But such a shortage could develop before long, pushing the price still higher for market reasons.

The prospect of this happening emerges from a close analysis of likely demand and supply trends.

As to demand, it seems probable that the rate of increase in the medium term will be at a slower rate than during the past couple of decades. This will largely be a consequence of the higher cost of energy. However, there seems little doubt that, in normal economic circumstances, energy demand will continue to grow, as is already evident in those industrialised countries which are coming out of the recent recession.

There are two further factors likely to create problems on the demand side. The first is the question of regional imbalance. As far ahead as we can see (in the present state of energy technology) the main industrialised areas of the world outside the communist bloc are likely to have to import a substantial part of their energy requirements. This will remain true in particular of Western Europe, the United States and Japan. So long as this situation remains it puts a very strong card, in both an economic and a political sense, in the hands of the energy suppliers.

The second factor is the rate at which the developing countries will industrialise and improve their living standards. As was shown in chapter 4, a relatively small increase in the per capita energy consumption of the developing countries in Asia, Africa and Latin America, coupled with the very large population increases in these regions, could have a major impact on world energy demand. Rightly, the developed countries are seeking to help the developing world; but to the extent that these policies succeed, other problems will be created, not least in the field of energy.

Overshadowing these uncertainties on the demand side, is the risk that, even at a very modest increase in the rate of demand, shortages could develop within the next decade or so in the supply of the most convenient fuels, namely oil and natural gas, and that other fuels will have great difficulty in filling the gap. This is certainly the conclusion of the most important recent studies of world energy prospects, such as the 1977 Reports of the OECD, the CIA (on which President Carter's energy strategy is largely based) and the Energy Workshop.

If it is assumed, and it seems a prudent assumption in the light of the most authoritative present estimates, that even with modest demand growth rates there could be real supply difficulties before the end of the century, leading to progressive increases in the real price of energy, what then should be the policy to be pursued? Before answering this question it is important to consider the steps that have already been taken in some of the main consuming areas.

Broadly speaking, it can be stated that there was a very vigorous reaction to the OPEC price increases in 1973 and 1974. In all areas dependent upon imported oil supplies major policy reviews were carried out. The main emphasis was placed on energy conservation, because a shortage of supplies appeared to be the immediate issue. Consideration was also given to developing alternative resources, indigenous if possible, to reduce import dependence and oil dependence in particular. As a longer term measure, it was proposed to devote larger resources to research, both to improve the production and use of existing fuels and to develop new ones.

However, four years after these events, it can equally truly be said that the effort has tailed off. This appears to have been due to two main factors. The first is the economic recession which emerged in 1975 and continued through into 1976 and 1977, with only limited signs of an upswing in the latter year. The second is the fact that consumers gradually became accustomed to the higher level of prices and as no evidence of shortage became apparent (largely because of the recession), the edge wore off the efforts which some governments and international organisations were seeking to stimulate. This was particularly true in the USA and the European Economic Community.

Coal and Energy

Reference has already been made in earlier chapters to the energy problem in the United States. Essentially it arises from the fact that American energy self-sufficiency began coming to an end at the very time that OPEC was becoming more powerful. America, in recent years, has become increasingly dependent upon oil and gas for its energy needs; indeed it was largely due to American initiative and enterprise that the industrialised world as a whole has moved so strongly in this direction. So it was understandably a matter of major concern when supplies of natural gas were insufficient to meet the rather more than seasonal cold weather of the winter of 1976/7; and when getting on for 50 per cent of the oil consumed had to be imported at the new and much higher world prices.

This explains the initiatives launched by recent Presidents, culminating in the very wide ranging measures submitted by President Carter to Congress in mid 1977. Essentially the object is to reduce oil import dependence from the present level of some 9 million barrels a day (over three times what is expected from the UK sector of the North Sea at its peak) to 6 million barrels a day by 1985; and to achieve this without interfering with continued growth in industrial production and living standards.

The question remains how effectively this ambitious energy plan can be carried out. There appear at the moment to be two major obstacles. The first is that there is a widespread public distrust of talk of future energy shortages, which is thought to be largely inspired by the oil companies in their own interests. In fact, the alarm signals have come much more from the American Administration than from any sectional interests; but the American consumer remains difficult to convince so long as he does not experience a real shortage. A prospective difficulty in making energy ends meet leaves him largely unmoved.

The second obstacle arises from the increasing contradiction between the efforts to develop and use more indigenous energy resources on the one hand, and to preserve the environment on the other. This is particularly apparent in the case of the United States' two long-term energy prospects, nuclear power and coal. Strong measures to safeguard the environment are, on present indications, going to make it very difficult to achieve the increase in production of these indigenous resources in sufficient quantity to restrict the flow of imported oil and ease the potential shortage of natural gas.

Thus the position in the United States is that, although the Administration has shown determination to deal with the energy problem, the lack of public response and the degree of environmental restraint, make the outcome questionable.

To some extent, similar inhibiting factors apply within the EEC.

The major difficulty there, however, is the variety of interests of the member countries, with Britain at one end of the scale potentially self-sufficient in energy, and Italy, Belgium, Luxembourg and Denmark at the other end largely dependent on imports. It is not surprising that, as the supply position has eased since 1974, the readiness of the EEC countries to commit resources on any joint basis to make themselves less energy dependent in the future has diminished. There have, however, been a number of initiatives within the Community aimed at developing a combined and dynamic energy policy. The Commission has put forward positive proposals with this in mind but, due to the differing interests of member countries, only a limited amount of practical progress has so far been made. The European Parliament, too, has been pressing for a more effective policy.

Among the coal industries of the Community a clearly defined policy for expanding the production and use of coal has been put forward. This has resulted from much work, conducted in a remarkable atmosphere of unanimity, among such leaders of the Community's coal industries as Dr. Karlheinz Bund, Chairman of the Executive Board of Ruhrkohle, Germany, Paul Gardent, Director-General of the Charbonnages de France and Marcel Peeters, President of the Belgian Fedechar, together with my colleagues and myself from the N.C.B. We have been ably assisted by Andre Woronoff the Secretary-General of our association (known as CEPCEO).

Among the institutions of the Community where coal policy can be discussed and recommendations made, is the Coal and Steel Consultative Committee which is located in Luxembourg. The President of this Committee in 1976/7 was Joe Gormley (the President of the NUM) under whose effective guidance a number of important debates took place both on coal and steel – the latter of which is going through a particularly difficult phase at present.

In spite of these and other efforts to bring about an effective and meaningful energy policy within the Community, the results up to the present have been limited. It may well be that until there is a further world crisis of the sort which occurred in 1973 and 1974 positive and concerted Community action in the field of energy may be difficult to achieve. But by then it may be too late to take some of the more fundamental measures. In the meantime it is important that the individual member countries of the Community should act vigorously on their own initiative. In this, I believe Britain has a crucially important role to play because of its substantial and varied energy resources.

Much has, indeed, already been achieved in Britain. The policy for the present exploitation of the oil and gas in the North Sea has been devised and applied, with growing benefits to the country's balance of payments. A plan for the development of the coal industry up to 1985 has also been agreed and is being carried out, with an extension of this plan to the year 2000 being actively considered by government. It is true that the long-term development of nuclear generation has yet to be decided, but, in the light of the present controversy on this subject and the availability of other resources, a period for further consideration may not be unreasonable.

What is now needed in Britain is to bring these various developments, or prospective developments, together in a single coherent strategy (which obviously would have to be sufficiently flexible to take account of future variables). This no doubt is what Tony Benn has in mind in setting up an Energy Commission on which are represented the producers, unions and users.

In view of the likelihood (based on most available evidence) of an increasing stringency in world energy supplies as the century draws to a close, and the high probability of significant increases in energy prices, it is fairly clear what the constituents of such a policy should be both in Britain and elsewhere.

The essential feature in the situation is to find and develop additional energy resources and make the best use of them. A four-point approach therefore seems called for:

Exploration: There are abundant reserves of energy, especially coal, in the ground. Before these can be exploited they must be identified by exploration. Since the energy crisis of 1973/4 exploration efforts have increased and have yielded positive results. They now need to be intensified.

Investment: A major allocation of resources has to be made to exploiting energy if prospective shortfalls are to be avoided. The high price of oil, and the prospect of further price increases, will provide new opportunities for economic investment in energy supply. Because of the long lead times involved – and coal is a prime example of this – major investment decisions have to be made many years before the product is likely to be required. In spite of all the uncertainties, this means taking a view on the longer term since inaction in the face of likely long term shortages could involve great risk. Investment is likely to be required not only in producing the energy, but also in transporting it in increasing quantities. This means more attention to infrastructure developments, such as railways, ports and ships. In the case of coal, more than twice as much is likely to have to be moved internationally by the year 2000 than is moved now.

Research: New methods of producing and using energy need to be devised. Among the conventional fuels this is particularly true of coal. Unless more automated and less man-intensive ways of producing coal are devised, one of the major impediments to expansion in the long-term could be lack of manpower. Equally, in coal usage, which is at present relatively less convenient than other fuels, big strides need to be made in conversion technology, and by other means to increase its convenience. Research in new sources of energy also needs to be stepped up; but as has been indicated in earlier chapters these are unlikely to have a major impact until well on in the next century.

Conservation: The present high price of energy, the likelihood of further real increases and of supply shortages makes a positive policy of conservation and greater efficiency in the use of energy of prime importance. Some progress has been made in some countries, such as the UK but much more determined efforts are required.

These, then, in my opinion are the four requirements of the energy strategy and programmes of action which need to be devised country by country and internationally where appropriate in the light of likely trends to the end of the century. A degree of hesitation and uncertainty has developed since the initial shock of the oil price increases of 1973 and 1974. It is a matter of common interest throughout the world, as amply demonstrated at the World Energy Conference held in Istanbul in 1977, that these doubts should be set aside and a positive plan of action put in train.

In that plan of action, coal, as the world's most abundant fossil fuel, has a major role to play.

Abbreviations

ACEC Advisory Council on Energy Conservation. (UK)
ACORD Advisory Committee on Research and Development. (UK)
AEA Atomic Energy Authority. (UK)
AGR Advanced Gas-Cooled Reactor.
ATM Advanced Technology Mining.
BCURA British Coal Utilisation Research Association.
BNOC British National Oil Corporation.
BP British Petroleum.
CEGB Central Electricity Generating Board.
CEPCEO Association of Coal Producers of the European Community.
CRE Coal Research Establishment. (UK)
DFR Dounreay Fast Reactor.
DOE Department of Energy. (US)
DOE Department of the Environment. (UK)
GW Gigawatt = 10^9 Watts.
ECSC European Coal and Steel Community.
ELSIE Electronic Signalling and Indicating Equipment.
EPA Environmental Protection Agency. (US)
ERDA Energy Research and Development Administration. (US)
ETSU Energy Technology Support Unit. (UK)
FEA Federal Energy Administration. (US)
HTR High Temperature Reactor.
IEA International Energy Agency.
JET Joint European Torus.
LNG Liquefied Natural Gas.
MRDE Mining Research and Development Establishment. (UK)
MW Megawatts = 10^6 Watts.
NCB National Coal Board.
NPC Nuclear Power Company. (UK)
NRDC National Research Development Corporation. (UK)
OCR Office of Coal Research. (US)
OCD Overseas Coal Developments. (UK)
OECD Organisation of Economic Co-operation and Development.
OPEC Organisation of Petroleum Exporting Countries.
PFA Pulverised Fuel Ash.
PFR Prototype Fast Reactor. (Dounreay)

PRT	Petroleum Revenue Tax. (UK)
PSA	Property Services Agency. (UK)
ROLF	Remotely Operated Longwall Face. (UK)
SGHWR	Steam Generating Heavy Water Reactor.
SNG	Substitute Natural Gas.
TCE	Tons of coal equivalent.
WAES	Workshop on Alternative Energy Strategies.
WAGR	Windscale Advanced Gas Reactor.
ZEBRA	Experimental reactor for studying fast reactor physics. Located at Winfrith, Dorset. (UK)

Glossary of Terms

Bunkers	In marine terminology, now used to refer to fuel, originally coal now usually oil.
Calorific Value (CV)	A measure of the amount of heat obtained from a fuel, expressed in British Thermal Units (BTU), therms, joules or multiples of joules.
Equity Oil	Oil which oil companies sell as their share in participtation agreements with the host country.
European Community	A term embracing the nine countries of the European Economic Community (EEC) – Belgium, Denmark, France, West Germany, Ireland, Italy, Luxembourg, Netherlands, United Kingdom.
Fluidised Bed	A technique by which coal is burned in a bed of finely divided particles of any suitable mineral matter through which air is passed.
Formed Coke	A process for forming low ranking coals into brickettes suitable for use in blast furnaces.
Fossil Fuel	A term applied to coal, oil and natural gas.
Landed Cost	The cost of oil in the importing country, including transport costs and oil company profits as well as the price paid to the producer.
Lignite	A lower grade of coal (that is, with a lower calorific value, *qv*). Also known as "brown coal".
Liquefaction	Turning a gas into a liquid, e.g. in the case of natural gas for easier transport.
Lurgi	A process for producing gas from coal originated by the German Lurgi comapny in the late 1920's.
Participation	A term applied in the oil industry to the agreements under which host countries shared with oil companies in the exploitation of their reserves.
Participation Oil	The share of the oil produced under a participation agreement which goes to the host government.
Posted Price	A notional price forming the basis on which host government royalties and taxes are calculated.
Primary Fuel	A fuel which can be used in the basic form in which it is obtained, such as coal, petroleum, natural gas, nuclear and hydro electricity.

Pyrolysis	Breaking down a substance by the application of heat. In the coal industry it can be used, *inter alia*, to yield substitute natural gas.
Secondary Fuel	A fuel which is derived from another source, e.g. electricity from coal, oil or natural gas.
Strip Mining	American equivalent of open-cast mining in Britain.
Telechirics	The control of operations by an operator located some distance away from the work, e.g. coal cutting at the face.

Chronological Table

1947	UK coal industry nationalised. National Coal Board established Jan. 1st.
1948	British Electricity Authority takes over former municipal and private undertakings.
1949	UK gas industry brought into public ownership. Gas Council established May 1st.
1954	UK Atomic Energy Authority established.
1956	Suez crisis.
1956	First UK Atomic Power Station, Calder Hall, officially opened (Oct. 17th).
1958	British Electricity Authority replaced by Electricity Council and Central Electricity Generating Board.
1959	Dutch Slochteren gas field discovered.
1960	Organisation of Petroleum Exporting Countries (OPEC) founded.
1964	First licences issued for North Sea oil and gas exploration.
1965	First UK North Sea gas strike (West Sole, October).
1967	Egypt-Israeli "Six Days" war; Suez Canal closed.
1967	First natural gas ashore from North Sea West Sole field.
1970	OPEC meeting at Caracas, Venezuela – first major increase in posted oil prices and general increase in tax rates.
1971	Teheran Agreement between Arabian Gulf oil producers and oil companies to raise posted prices and host government "take".
1971	Tripoli Agreement between Libya and oil companies to raise posted prices and government "take".
1972	Geneva Agreement raising posted prices to offset dollar devaluation.
1972	Gas Act in UK restructures the nationalised gas industry.
1973	First participation agreements concluded between oil companies and host governments.
1973	Middle East ("October") War.
1973	Arabian Gulf oil producers unilaterally raise prices by 70 per cent.
1973	Oil supplies restricted if importers considered to be unsympathetic to Arab cause, including total embargo on supplies to USA and Netherlands.

1974	Biggest posted price rises of all, resulting in five-fold increase in landed crude oil prices in previous three years.
1974	Establishment of the International Energy Agency (IEA).
1974	Department of Energy established in the UK.
1974	Plan for Coal issued, by the National Coal Board in agreement with Government and mining unions.
1975	First North Sea oil ashore from Argyll Field (June).
1975	Oil Taxation Act introduces the Petroleum Revenue Tax (PRT) to secure reasonable share of offshore oil revenue for the nation.
1975	Petroleum and Submarine Pipelines Act includes provision for the establishment of the British National Oil Corporation (BNOC).
1975	Oil production starts from the Forties Field (Nov.).
1976	Oil production starts from five North Sea fields – Auk (Feb.), Beryl (June), Montrose (June), Brent (Nov.) and Piper (Dec.).
1977	Plan 2000 announced by National Coal Board (Feb.).
1977	President Carter's energy policy announcement (April).
1977	Tenth World Energy Conference Istanbul (Sept.).

Index

Printed in Great Britain by
Lewis Reprints Ltd,
Tonbridge, Kent